essential atlas
of physical
geography

BARRON'S

First English-language edition for the United States, Canada, its territories and possessions published in 2003 by Barron's Educational Series, Inc.

English-language edition © Copyright 2003 by Barron's Educational Series, Inc.

Original title of this book in Spanish: *Atlas Básico de Geografía Física*

© Copyright 2002 by Parramón Ediciones, S.A., World Rights

Published by Parramón Ediciones, S.A., Barcelona, Spain

Authors: Parramon's Editorial Team

Illustrations: Parramon's Editorial Team

Text: José Tola

Translation from Spanish: Eric A. Bye

All inquiries should be addressed to:
Barron's Educational Series, Inc.
250 Wireless Boulevard
Hauppauge, New York 11788
http://www.barronseduc.com

International Standard Book Number 0-7641-2511-7

Library of Congress Control Number 2003101098

Printed in Spain

9 8 7 6 5 4 3 2 1

PREFACE

This *Essential Atlas of Physical Geography* places in the reader's hands a wonderful opportunity to learn about the formation and evolution of our planet, as well as the characteristics of and transformations undergone by the features that make up the landscape. It is thus an extremely useful tool for exploring the wonderful home that shelters the living beings represented by plants and animals, whose relationship with the environment is an essential part of the Earth's ecological balance.

The various sections of this book constitute a complete synthesis of geology. They contain many illustrations and charts that show in detail the characteristics of minerals, the planet's activity, the shape of the landscape, the climate, and the importance of water both in the seas and on land. These illustrations, which are the nucleus of the book, are complemented by brief explanations and notes that make it easier to understand the main concepts, as well as a complete subject index that makes it easy to find any topic of interest.

Our goal in creating this *Essential Atlas of Physical Geography* was to produce a practical and instructional work that is useful, accessible, scientifically accurate, and at the same time, interesting and clear. We hope the reader will agree that we have achieved our goals.

CONTENTS

INTRODUCTION

PHYSICAL GEOGRAPHY

The science of physical geography is devoted to the study of physical phenomena that take place on the surface of the **earth** and produce everything we see around us that we refer to as landscape. This entire series of phenomena affects the planet's three major environments—land (which makes up the **lithosphere**), air (which constitutes the **atmosphere**), and water (which is the **hydrosphere**).

STUDYING LANDSCAPES

The geographical features that dominate the landscape, such as mountains, plains, and valleys, strike us as eternal and unchangeable, and most of the time we are not able to observe changes of any importance. Just the same, these features are in continual transformation, even if the speed at which this happens is too slow for us to perceive and the process takes centuries or even several million years to complete.

We speak in terms of recent **mountains**, which have jagged peaks and are tremendous in size; they

Even though we cannot see it, the ocean floor of our planet contains considerable high relief.

are the result of occurrences that took place before our species even existed. The life span of any organism is very short compared to the scale on which **geological processes** take place. The longest-living organisms are certain trees that can live as long as 3,000 years, but these three millennia are just a brief passage in the geological life of the planet.

EXPLORATION AND DISCOVERY OF THE EARTH

Physical geography was initially limited to conventional **geography**, and the first geographers described the features of the earth's crust as changeless elements that defined locations and were used as reference points. In those early days people debated whether the earth was flat or spherical, because its curvature was imperceptible from a human point of view. For a long time it was supposed that we lived on a great extension of solid ground that floated on a huge sea but ended in a great void. The navigators who crossed the **Mediterranean** in antiquity didn't dare to go farther than the "columns of Hercules," the present Straits of Gibraltar, unless they could keep the coast in view. They believed that beyond the horizon the sea

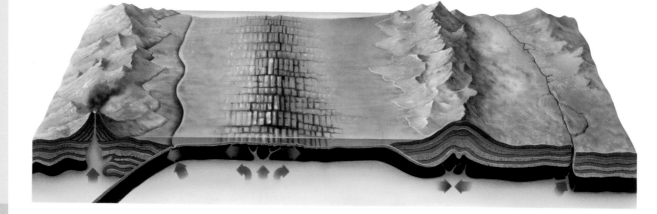

An oasis is a portion of terrain that offers conditions for plant life and human activity in the middle of a desert area.

would end suddenly in a tremendous waterfall that would pull them in irretrievably, causing the destruction of their ships and the death of the entire crew.

Even though as early as the **classical Greek era** mathematicians had made calculations in various parts of the known world that allowed them to assert that our planet was a sphere, the idea that it was flat persisted well into the Middle Ages. At that time the first **maps of the world** started to appear and showed a more accurate representation of the continents.

When **Columbus** undertook his voyage from Palos de Moguer (in southeastern Spain) in command of three ships outfitted by the Spanish crown, he hoped to reach India and the countries that produced Asian spices by following an alternative route toward the west because the commercial routes toward the east were under the control of other European powers. However, this would be possible only if one believed that the Earth was round, and there were many people at the time who were still doubtful. Even so, Columbus had information from other navigators and was convinced that it was indeed possible to reach Asia by circumnavigating the globe toward the west; but what he didn't know was that between Europe and Asia there was another continent that would later be named America.

Years later, in 1519, Magellan left Spain to sail around the world, and although he died in 1521 in the Philippines, his crew, under the command of Juan Sebastian Elcano, managed to return to Seville six months later, thereby completing the first circumnavigation of the globe. After that, the great journeys of discovery made by the maritime powers, in

which scientists and geographers began to participate regularly, made it possible to formulate a more precise idea of what the world was really like. The data from every voyage answered many questions and made it possible to draw an accurate picture of the **exposed landmasses** surrounded by a great expanse of water that together make up the planet Earth.

As early as the first half of the eighteenth century geography was taking on great importance and becoming a modern science. It became an established science after a little more than a century, and in the middle of the nineteenth century it was possible to speak of **physical geography** as we know it today, with the collaboration of two other sciences, **physics** and **chemistry**.

The landscape is shaped by internal forces (earthquakes, plate tectonics, faults, and so on) and by external agents (wind, rain, ice, etc.).

7

A GLIMPSE OF OUR PLANET

In this atlas we will see the broad fields with which physical geography is concerned that have encompassed the entire planet from the time of its formation up to the present.

Seen from space, the **Earth** is a spinning sphere, which, along with the other planets, revolves around the sun. Its color has varied throughout the millions of years of its existence because of the atmosphere that surrounds it. In order to understand the present landscape, we have to go back to its origins. How did the Earth originate? It appeared along with the other planets and the **sun** when the **solar system** was formed. By studying its composition we can determine what it is made of, and we can do the same with the atmosphere because its activity is very important in landscape formation.

Minerals and **rocks** are one essential feature involved in getting to know our planet. They were formed at the outset, and since that time they have undergone changes to produce the variety that we know today. To understand these changes better, the geological history of the Earth has been divided into major **periods**, each of which has its own characteristics.

When the material that makes up geographical features is considered in this light, we can see the planet's history, evolution, and dynamism.

A LIVING PLANET

At one point the Earth's crust solidified and produced an environment where the presence of water made it possible for **life** to appear. However, the crust is a fairly thin layer that could be compared to the skin of an orange. Underneath the planet is still a molten mass, just as it was at the beginning. There are currents of matter inside this mass, and in some places the hot mass breaks through the thin, solid covering and emerges to form new rock. This process is known as **volcanism**.

There are **volcanoes** both on the **continents** and on the **ocean** floors, and they contribute to the formation of new crust. They also form new material on the lower part of the **plates** that make up this cortex, even though in other places these plates sink down and melt at the same time so that there is not much variation in the total mass.

These events are part of the **internal activity** of the planet. This activity is responsible for formation of the large structures that make up the Earth because that

With time, water can carve deep, beautiful defiles, such as the Gela Canyon (Utah).

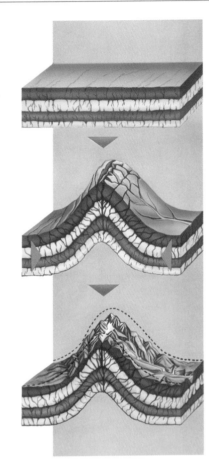

Geomorphology is the science that makes it possible to study the formation of the Earth's relief.

is how such features as continents, islands, and mountains are formed and disappear. However, the appearance of these features today is different from the way they looked when they were created because even the gaseous layer that surrounds us is an environment undergoing constant evolution. When the air moves, it produces **wind**, and it carries water vapor that under the right conditions can condense and fall in the form of **rain** or **snow**, depending on the temperature. The activity of these external agents changes the earth's crust—in other words, they transform it through erosion.

Water plays a crucial role in all these processes. Therefore we will study it in all of its aspects. Most of the water on the planet is found in the **oceans** and **seas**, constituting a tremendous mass that covers two-thirds of the Earth's surface. When the level of the water changes, and when the tremendous forces of currents, tides, and waves come into play, the ocean keeps the appearance of the continents in a continuous state of change. There are also bodies of water inside the solid ground, but they are much

smaller compared to what exists in the oceans; still, they are extremely important in the formation of landscapes. In addition, the effects produced by the continental water masses take place at greater speed and are often perceptible in a single generation.

Physical geography studies our planet as we see it today and not just geological forces and other phenomena from the past. To complete the overview of this science, we will devote a few pages to the way landscapes are formed, in other words, to **erosion**. Water, wind, and ice all continually wear down the rocky surface in a slow but continuous process.

As a result of this ongoing work, the rough shapes produced by the inner forces become smoother and the surface becomes covered with a thin layer that we know as **soil**.

Finally, as a practical complement, we will see how humans make representations of the seas and continents in a visible, approachable size by creating **maps** on different scales and for different purposes.

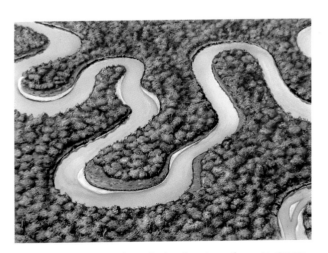

Meanders are formed in nearly flat terrain where the water creates a series of curves as it seeks out the easiest soil to erode.

THE FORMATION OF THE EARTH

The Earth that we know today looks very different than when it originated about 4½ billion years ago. At that time it was a mass of rocks thrown together, the interior of which heated up and melted the entire planet. Later on, though, another solid crust formed, a gaseous layer appeared, and water was formed. These two elements began to change the shape of the crust produced by the planet's internal activity.

THE ORIGIN OF THE UNIVERSE

According to astronomers' calculations, about 15 billion years ago there was an explosion of unimaginable size. A mass of tremendous density, in which the atoms were packed tightly together, exploded. This event is referred to as the **big bang**. The force it unleashed scattered the dense material in all directions at a speed approaching that of light.

With time, as the masses of this material got farther away from the center of the explosion and slowed down, material that was close together formed **galaxies**.

According to the most widely accepted theory, the origin of the universe was a great initial explosion known as the big bang.

ATOMS

Atoms are the basic units of all materials. They consist of a nucleus that contains neutrons and protons surrounded by spinning electrons.

 Galaxies are clusters of up to a billion stars; they have diameters of several thousand light-years.

THE RIGHT DISTANCE

The Earth is about 93 million mi (150 million km) away from the sun—just the right distance for maintaining an atmosphere and keeping water in a liquid state.

THE FORMATION OF THE SUN AND ITS PLANETS

The **Milky Way** condensed as a fairly dense cloud from a quantity of stellar material near the edge of one of the many galaxies produced in the great explosion. The forces of **gravitational attraction** caused most of this mass to form a large central sphere with smaller masses spinning around it. The central mass turned into a large molten sphere, a **star** that we know as the **sun**. The smaller masses that revolved around it became the **planets**. All of these bodies together make up the **solar system**.

 The moon has no atmosphere and is exposed to a continual bombardment by meteorites, even though most of them are small.

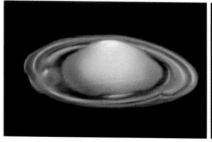 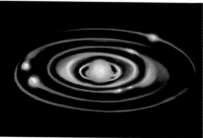 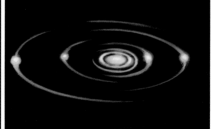

The phases in the formation of the solar system. Starting with interstellar material, a star (the sun) condensed in the middle, and other celestial bodies (the planets) condensed around it and remained in rotation.

THE SOLID PLANET

When the sun was formed about 5 billion years ago, the **Earth** was also formed. After an initial period when it was a molten mass, the outer layers began to solidify, even though the heat coming from the interior melted them again. Finally the temperature became low enough to allow the formation of a stable **crust**. At first there was no atmosphere, and it was continually bombarded by meteorites. The volcanic activity was also intense, and great masses of **magma** were released to the surface, thereby increasing the thickness of the crust when it cooled and solidified.

Venus

A PROTECTIVE SHIELD

The atmosphere is a covering that protects against radiation and most meteorites from outer space.

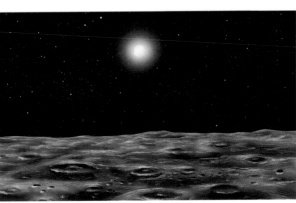

Mars

The earth's crust was formed many times before it finally solidified.

THE SEAS AND THE ATMOSPHERE

Volcanic activity generated a great quantity of gases that formed a covering around the Earth. Their composition was very different than it is now, but this covering was the first protective layer that made the appearance of water possible. **Water vapor** was generated from oxygen and hydrogen in volcanic eruptions, and the vapor condensed to form **rain**. Millions of years of rain added up to produce a covering of water, the **hydrosphere**.

Mercury

The Earth

The present atmosphere is the result of activity by organisms that inhabited the oceans after life first appeared.

Atmospheric conditions on Earth make life possible. Venus has a dense atmosphere, Mars has a thin one, and Mercury has none at all.

THE COMPOSITION OF THE EARTH

If we were to cut through our planet, we would find different layers beneath the crust, and there is considerable variation in their structure. They are all superimposed on and wrapped around one another.

Outside the crust there are two other layers that are very important in the study of physical geography: the hydrosphere and the atmosphere.

THE LAYERS OF THE GLOBE

The Earth is not a uniform rocky sphere but rather is made up of very diverse materials that are found in different states and are distributed in characteristic ways. The data that geologists have gathered indicate that if we were to cut through the surface and into the center, we would find three basic layers: the thin **crust**, the **mantle** beneath it, and finally the **core**. But the crust in turn contains a superficial layer (**sial**) and a deeper one (**sima**); the mantle consists of an outer and an inner part and even the core is divided into outer and inner parts.

The Earth's magnetic field affects more than the surface, for it extends thousands of miles (kilometers) into outer space.

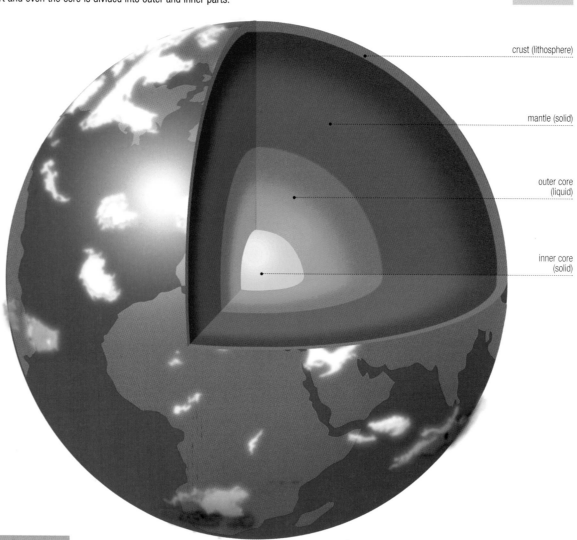

crust (lithosphere)

mantle (solid)

outer core (liquid)

inner core (solid)

DISCONTINUITIES

The core is separated from the mantle by an area where there is an abrupt change in the density of the rock known as the **discontinuity of Mohorovicic** (known as Moho). The inner mantle is separated from the outer core by another area with another abrupt change in density: **Gutenberg's discontinuity**.

THE CORE

The innermost layer of the planet is not a dense, solid mass, as we might think; rather, it is in a semi-liquid state in spite of the tremendous pressure that exists. Of the two parts that make it up, the inner one is the denser; it is mostly made up of **iron** and a small percentage of **nickel**. On top of that, enclosing it all, is the outer core whose composition is more complex and totally unknown. But evidently it consists of oxygen, carbon, sulfur, hydrogen, and potassium, and it is less dense than the inner core.

STUDYING THE INTERIOR OF THE EARTH

Geologists have succeeded in penetrating only the most superficial part of the planet, the crust, and have not drilled all the way through it because of insurmountable technical difficulties. But there is an indirect way to determine the composition of the interior: It involves sending **vibrations**, whether natural (**seismic**) or artificial (**explosions**). By measuring the speed at which the vibrations move, it is possible to determine what type of material they are moving through and how thick it is.

THE MAIN COMPONENTS OF THE CRUST

Element	Continental Crust (%)	Ocean Floor (%)
aluminum	16	16
calcium	5.7	11.1
iron	6.5	9.4
magnesium	3.1	8.5
potassium	2.9	0.26
silicon	62	49
sodium	3.1	2.7
titanium	0.8	1.4

INTERNAL HEAT

It is believed that about one-third of the planet's internal heat comes from **thermal energy** accumulated during its formation, and that it is slowly escaping. The rest is due to the decay of **radioactive elements**. The temperature of the core is about 8,492°F (4,700°C), and the mantle probably does not exceed 1,832°F (1,000°C), for otherwise the rock would melt, and we know that it's solid. Finally, the crust increases in temperature about 86°F (30°C) for every kilometer of depth. But half the heat that escapes from the crust is heat produced through radioactivity.

Using drill rigs, geologists can extract samples of the subsoil and study its makeup.

THICKNESS OF THE EARTH'S LAYERS

Layer	Approximate thickness
crust	4–43 mi
upper mantle	415 mi
lower mantle	1,383 mi
outer core	1,376 mi
inner core	775 mi

When a wave goes through two mediums of different density, it experiences a deviation that depends on the density of the materials.

Iron is one of the main components in soil. The photo shows an open iron mine.

THE ATMOSPHERE

The Earth's outer layer is gaseous and of a very different composition and density than the solid layers below, but this is the area in which life exists, and it is also very important in the erosion processes that have produced the current landscapes. The changes that occur in the atmosphere have an important effect on these processes.

THE ORIGIN OF THE ATMOSPHERE

When the planet was formed, the lightest elements in a gaseous state remained on the outside and produced a layer, but they probably disappeared shortly thereafter. When the crust solidified, the intense **volcanic activity** that took place initially caused the formation of numerous denser gases that made up the **primitive atmosphere**. Its composition was very different from that of the atmosphere today, and it was probably made up of **carbon dioxide**, **sulfur dioxide**, and **water vapor**. After life appeared, the activities of organisms that produce **oxygen** caused the concentration of this element in the atmosphere to increase until it reached the present composition.

The atmosphere makes life possible on our planet, and some of its products (water, ice, wind, etc.) contribute to landscape formation.

Fog is a stratified cloud (made up of tiny water droplets) that comes down to ground level.

The atmosphere acquired its present composition about 370 million years ago.

About 2 billion years ago the atmosphere contained only 1 percent oxygen.

AN EXTREMELY VARIED MIXTURE

In addition to the elements and chemical compounds listed in the tables, the atmosphere contains many other components. One very important one is **water vapor**, which has an effect on a region's climate and on the existence of plants. It also contains very tiny solid particles in suspension, such as sand from deserts, ashes from volcanic eruptions, plant pollen, and even small organisms that make up a type of **aerial plankton**. All these factors have an effect on the atmosphere's clarity and many associated phenomena such as fog, rain, and others.

APPROXIMATE COMPOSITION OF DRY AIR

Component	Percent of Volume	Percent of Weight
nitrogen	78	75.58
oxygen	20.95	23.16
argon	0.93	1.28
carbon dioxide	0.035	0.053
other inert gases	0.0024	0.0017
hydrogen	0.00005	0.000004

Rain is precipitated water droplets formed from water vapor present in the atmosphere.

THE LAYERS OF THE ATMOSPHERE

The atmosphere extends from the surface of the Earth's crust up to an altitude of more than 1,240 mi (2,000 km), but it becomes progressively thinner. The following layers are distinguished, from the surface to outer space: the **troposphere** (the densest layer, in which life occurs; it extends out to 5 to 11 mi [8–18 km]); the **stratosphere** (no clouds are formed, and the air is thinner than in the troposphere; it extends out to 31 mi [50 km]); the **mesosphere** (out to 53 mi [85 km]); the **ionosphere** (out to 248 mi [400 km]); and the **exosphere** (from 248 mi [400 km] to outer space; however, there is practically no more atmosphere at the level of 1,240 mi [2,000 km]).

CHARACTERISTICS OF AIR AT DIFFERENT ALTITUDES

Altitude Above Sea Level (ft)	Pressure (lbs/in²)	Temperature (°F)
0	15	59
1,640	15	52
3,280	13	47
4,920	13	42
6,560	12	36
9,840	11	24
13,120	9	12
16,400	8	1
24,600	6	−27
32,800	4	−58
49,200	2	−69

exosphere (248 to 1,240 mi [400–2,000 km])

INCREASINGLY COLD

In the troposphere, the temperature drops an average of 6° for every kilometer of altitude.

ionosphere (out to 248 mi [400 km])

mesosphere (out to 53 mi [85 km])

stratosphere (out to 31 mi [50 km])

troposphere (out to 5–11 miles [8–18 km])

Only the first few kilometers of the atmosphere have enough oxygen to support life, and the amount of oxygen decreases with altitude.

GEOLOGICAL HISTORY: THE PRECAMBRIAN ERA

The Earth has undergone a great many changes since the time of its formation. The first stages, from the solidification of the molten mass until the appearance of a solid crust, have left scarcely any evidence of their existence because many of the rocks that were produced later melted. They still belonged to a very dark, primitive time for science, and it was only at the end of this period that lasting structures appeared that still survive.

THE HISTORY OF THE PLANET IN TIME

The 4½ to 5 billion years that the Earth has existed, according to different estimates, represent a very long period of time, and its progress is one of the major topics of study by geologists. However, the precise knowledge available is limited to more recent times, scarcely more than 500 million years. Earlier history, which includes that of most of the entire existence of the planet, contains fairly broad gaps and long periods about which nothing more than hypotheses can be formulated. This time in the geological history of the Earth is generally called the **Precambrian era**. This name indicates that it precedes the **Cambrian era**, which is when many **multicelled animals** begin to appear.

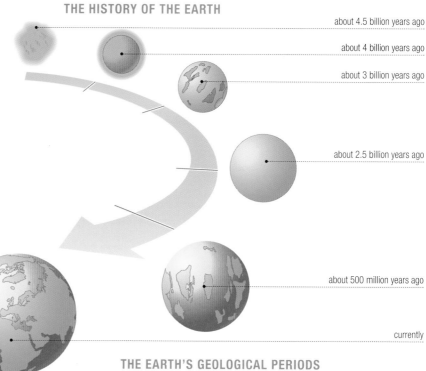

THE HISTORY OF THE EARTH

about 4.5 billion years ago

about 4 billion years ago

about 3 billion years ago

about 2.5 billion years ago

about 500 million years ago

currently

AN EON

An eon is the largest unit of geological time and is divided into various eras. Each era consists of various periods, which are divided into epochs.

The more recent a geological period is, the more information there is about it; therefore it is necessary to divide it into smaller sections.

PALEONTOLOGY

The science that studies fossils (the remains of living beings that inhabited the earth millions of years ago) is called paleontology. It is closely associated with geology.

THE EARTH'S GEOLOGICAL PERIODS

	Eon	Era	Period	Epoch	Millions of Years Ago
Precambrian	Hadean				4,500–5,000
	Archean				3,900
	Proterozoic				2,500
	Phanerozoic	Paleozoic	Cambrian		540
			Ordovician		500
			Silurian		430
			Devonian		410
			Carboniferous		345
			Permian		310
		Mesozoic	Triassic		225
			Jurassic		210
			Cretaceous		150
		Cenozoic	Paleogene	Paleocene	65
				Eocene	57
				Oligocene	34
			Neogene	Miocene	23
				Pliocene	5
			Quaternary	Pleistocene	1.6
				Holocene	0.01 (about 10,000 years)

DATING METHODS

The processes that geologists use to determine the age of rocks and minerals are called dating methods. The best data are obtained from procedures known as **radiometric determination**. One of them is called the **lead method**. It is based on the fact that radioactive elements such as uranium and thorium disintegrate until they turn into lead. It takes a certain amount of time for this to happen, and this is a known quantity. If uranium and lead isotopes are present in a mineral, it is possible to calculate the proportions of the two and determine how much

uranium has disintegrated, in other words, how much time has passed since the rock was formed. Other isotopes frequently used in this method are **rubidium-86** to **strontium-87** and **potassium-40** to **argon-40**, but the latter is less reliable because argon disappears quickly from minerals.

Galena is the only lead mineral found in large quantities in nature.

Carbon dating can be used to calculate ages up to 40,000 years; it is used primarily with human remains.

Various samples of marine fossils.

THE PRECAMBRIAN ERA

The Precambrian era, a long period in the Earth's history, lasted from the time of its formation up to about 540 million years ago and is divided into three eons in which great changes took place on the Earth. The **Hadean** is the oldest and includes all the time during which the Earth was mainly a red-hot sphere; it ended about 3.9 billion years ago when the crust stabilized. Then the **Archean** began; it can be considered the oldest period of the Earth after it solidified. During this time the crust kept cooling down, igneous and metamorphic rocks formed, and rains filled up the oceans. Then life appeared, about 2.5 billion years ago. This moment marked the start of the third eon, the **Proterozoic**. Until about 540 million years ago, life consisted of nothing more than one-celled animals. When the first multicelled organisms appeared, the Proterozoic eon and the Precambrian era ended.

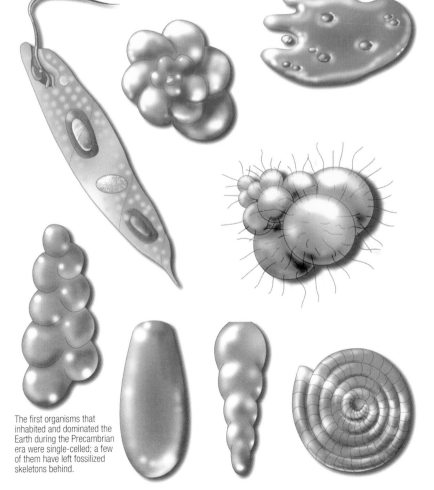

The first organisms that inhabited and dominated the Earth during the Precambrian era were single-celled; a few of them have left fossilized skeletons behind.

Proterozoic means "the time of first life" and indicates that during that time life was unicellular.

GEOLOGICAL HISTORY: THE PHANEROZOIC ERA

During the obscure time of the Precambrian era, most of the basic material that makes up the Earth's crust was formed and this was when the geological phenomena that affect us most directly took place. The Phanerozoic era began at this time; one of its characteristics is that it produced many fossils that show the presence of multicelled life on a planet that was already perfectly suited to living organisms.

THE PALEOZOIC ERA

The Paleozoic era lasted about 315 million years. The planet was very different from the way it is now. The exposed land looked like islands of various sizes scattered around the equator. Some of the main ones were **Laurentia** and **Gondwana**. Many geological folds were created during that epoch, and the main rocks were quartzites, shale, limestone, and arsenic. The climate was hot and humid. Multicelled animals began to spread and evolve, and life in the ocean became very diversified. The first plants colonized the land and soon formed large forests; echinoderms and, later on, fish appeared in the seas. Later still, amphibians, insects, and the first reptiles appeared.

Trilobites, a class of extinct arthropods, grew as long as 27 in. (70 cm). They were very abundant during the Paleozoic era and have come to symbolize it.

 In geological terms, extinction of the dinosaurs occurred quickly because it took just a few thousand years; this event was probably the result of a climatic change.

The great forests of the carboniferous period produced today's coal deposits.

THE MESOZOIC ERA

The Mesozoic era lasted about 160 million years. At the beginning, all the small continents that had existed during the preceding epoch had come together to form a single huge continent known as **Pangea**. The main geological folds appeared in the Americas, resulting in the Rockies and the Andes. The climate was hot, but drier than during the Paleozoic era. On solid land this epoch was dominated by **conifers**. Among the animals, reptiles became highly developed, and during the Jurassic period, they developed into **dinosaurs**. Before dinosaurs became extinct, birds and the first mammals appeared.

Scene from the Mesozoic era.

THE CENOZOIC ERA

The last and most recent era, the Cenozoic, lasted about 65 million years. The continents already looked much as they do now, but the **Atlantic Ocean** was still narrow because the coasts of America on the west and Europe and Africa on the east were still fairly close together. Also, **India** was in the middle of the Indian Ocean and moving toward Asia. Mountain formation was very active, and the large chains such as the **Alps**, the **Atlas**, and the **Himalayas** were formed. The climate gradually cooled off until large surfaces of the planet became covered with ice at the start of the **Quarternary period**. These ice sheets were called **glaciers**.

The mammoth, a type of elephant, was able to survive the periods of glaciation because of its thick skin and heavy coat. It became extinct a little more than 10,000 years ago.

Distribution of the continents 520 million years ago (top) and 80 million years ago (bottom).

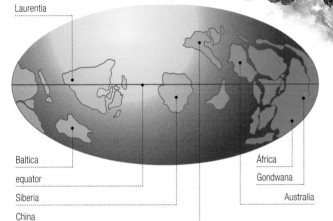

Laurentia

Baltica

equator

Siberia

China

África

Gondwana

Australia

equator

South America

North America

Antarctica

Australia

Asia

India

Africa

Within about 200 million years we will be able to travel directly between continents because they will have fused together again.

THE QUARTERNARY

The period from 2 million years ago to the present is designated the Quarternary period.

CONTINENTAL DRIFT

Continental drift is the name for the phenomenon by which the continents have moved throughout millions of years of geological history. It occurs when material emerges from the **mantle** beneath the ocean floor and creates a force that pushes on the areas occupied by the continents (the continental plates); as a result the continents change position. In addition, it often happens that part of a continent crumbles along one edge, and the result is a change in shape.

The Himalayas (which means "the dwelling of the snows" in Sanskrit) is a mountain chain that resulted from the collision of the Indian continental block and the Eurasian block.

CRYSTALLOGRAPHY

Collecting minerals is a very popular hobby—especially minerals that have certain shapes and exceptional beauty, many of which are designated as crystals. Minerals are one of the forms in which inorganic material occurs in nature. Many minerals have a characteristic geometric structure. Crystallography is the study of crystalline structures.

CRYSTALS

Minerals can appear in nature in two ways: without any particular shape (**amorphous**) or arranged in a very clear geometric structure (**crystals or crystalline minerals**). In order for crystals to form they need a certain amount of space; thus they commonly appear in cracks or hollows in rocks, or contained inside soft, malleable rocks that allow their growth.

pyrite

calcite

rock salt

ANISOTROPIES AND ISOTROPES

Many crystals react to a physical action in different ways depending on the direction in which it takes place; they are said to be **anisotropies**. Amorphous minerals are **isotropes**: They always react to a physical action in the same way regardless of the direction.

A **dihedral angle** is the angle formed by two indefinite planes that intersect one another.

A UNIVERSAL LAW

Crystals, as we have said, are characterized by having a specific geometric shape. However, when we simply look at a mineral, we find just a few geometric faces that are sometimes only half-formed and seem to have no specific shape. Even though on the outside they may appear to be irregular or partly misshapen, they are considered **regular crystals** in crystallography because they comply with a fundamental law: When temperature conditions are the same, crystals of the same type have the same dihedral angles. This is known as the **law of constancy of dihedral angles.**

The glass in a window is a melted, amorphous mass. Despite its solid appearance, it would melt down to a puddle in a few thousand years.

CRYSTAL CLASSIFICATION

In any crystal there can be three elements of symmetry:

1) **an axis of symmetry**: the straight line around which the crystal spins occupies the same position one or more times;

2) **a plane of symmetry**: the plane that divides the crystal into two symmetric halves;

3) **a center of symmetry**: a point where all the faces of the crystal are parallel by pairs.

Some crystals have all three of these elements of symmetry, others lack one of them, and a few lack all three. This is a basis for classifying all crystals found in nature: There are 32 **classes**, 31 of which have some type of symmetry and one of which has none.

Olivin, a rock of volcanic origin, is made up of rhomboid crystals.

THE THREE ELEMENTS OF SYMMETRY USED IN CLASSIFYING CRYSTALS

axis of symmetry

plane of symmetry

center of symmetry

SOME SIMPLE CRYSTALLINE SHAPES

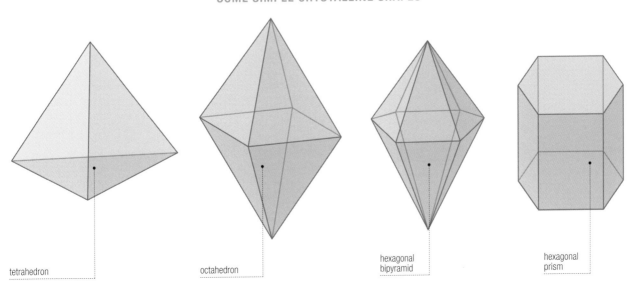

tetrahedron

octahedron

hexagonal bipyramid

hexagonal prism

MINERALS

Mineralogy is the science of minerals. Minerals are bodies of solid matter that occur in many different shapes, whether in isolation or as components of rocks. In order to study minerals we can consider their different properties, such as their hardness, their geometry (as we have already seen with crystallography), their chemical composition, and so forth. Minerals are very important because they are the source of many products that we use in our daily lives.

CHARACTERISTICS OF MINERALS

The glass in a window is not a crystal even though it is made using crystalline minerals, and a rock is also not a mineral but rather a material made up of several minerals. In order to understand what a mineral is, it is useful to consider its characteristics:

1) It is a **solid** substance because liquids have no fixed geometric structure.

2) By nature it is **inorganic**; thus, the shell of a mollusk is not a mineral even though it contains minerals.

3) It is found in **nature**.

4) It has an **established chemical composition** even though a contaminant may be present that alters its color.

MACLA

A macla is a combination of two or more crystals that retain their symmetry. Maclas can have very interesting, beautiful shapes.

Many times minerals occur in nature as masses inside a rock. This is known as a **vein** or a **seam**.

Oligist, also known as hematite, is a mineral very useful to humans because it produces very pure iron.

Block of pyrite macla.

Pieces of worked flint used by prehistoric humans for hunting, dressing hides, and so forth.

In prehistoric times people used minerals for making utensils, tools, and weapons.

THE MOHS HARDNESS SCALE

No.	Example	Characteristic
1	talc	easy to scratch with a fingernail
2	gypsum	can be scratched with a fingernail
3	calcite	can be scratched with a copper coin
4	fluorite	can be scratched with steel
5	apatite	can be scratched with steel
6	feldspar	barely scratches glass
7	quartz	easily scratches glass
8	topaz	easily scratches glass
9	corundum	easily scratches glass
10	diamond	easily scratches glass, diamond-scratchable only

THE PHYSICAL PROPERTIES OF MINERALS

In classifying minerals it is important to observe a series of physical properties:

1) **color:** because of impurities, some minerals show variations in color;

2) **color of the pulverized mineral:** scratching with a harder object produces a powder with a characteristic color;

3) **shine:** can be metallic (e.g., iron) or nonmetallic (e.g., nacreous or silky);

4) **refraction index:** a ray of light passing through a crystal is deflected at an angle that is characteristic of each mineral;

5) **birefringence:** some minerals divide a ray of light passing through them into two parts;

6) **luminescence:** some minerals emit light when they are illuminated.

Two minerals can have the same chemical composition but a different crystalline structure, such as diamond and graphite.

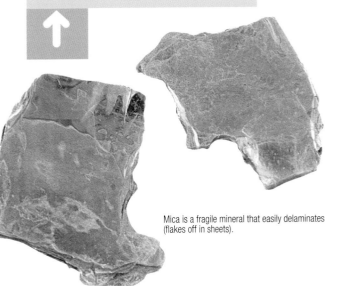

Mica is a fragile mineral that easily delaminates (flakes off in sheets).

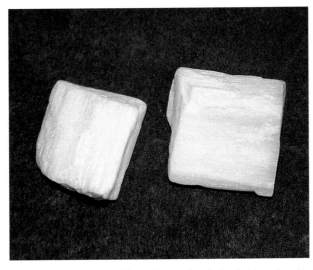

A diamond is pure carbon; it crystallizes in a cubical pattern, has great brilliance, and is extremely hard. The most beautiful specimens are highly valued as jewelry, but other pieces are used in industrial processes such as cutting, rectifying, and sharpening.

Gypsum, which occurs in many different shapes and is quite abundant, is much used in construction.

SOME OF THE MOST COMMON CRYSTALLINE MINERALS

Hardness	Name	Characteristic
1	talc	white, greenish, brown, or colorless; does not appear in isolated crystals
2	gypsum	white; sometimes forms large crystals
2	mica	generally colorless; forms leafy crystals
2.5	chlorite	greenish color, very small crystals that are scarcely visible
3	calcite	normally white, sometimes forms large crystals
3–4	serpentine	very different shapes and colors, sometimes fibrous like asbestos
4	fluorite	colorless, orange, or purple; forms cube-shaped crystals
5.5	amphibole	dark color, sometimes greenish; forms elongated crystals
6	apatite	variable color, grayish-white, bluish, or purple; forms many types of crystals
6	feldspar	white or slightly pink; forms maclas
6.5	olivin	olive-green, light or dark; forms various types of crystals
7	garnet	red, green, or brown; forms large crystals
7	quartz	white or clear, often forms large crystals

TYPES OF MINERALS

The minerals that make up the Earth's crust were formed from the original elements as a result of reactions that took place inside the planet. Therefore, there are a great many combinations.

In classifying them, they are grouped according to the way in which they were formed, by their crystallographic characteristics, by their chemical composition, and by other factors. We will now consider some examples of the best-known minerals.

THE CLASSIFICATION OF MINERALS BY ORIGIN

Although classification by origin is the least precise way of categorizing minerals, it is still useful in providing an initial idea of mineral classifications. Depending on the way in which they were produced, minerals can be

1) **magmatic minerals:** those that crystallized directly from magma. They have a clearly defined crystalline structure (e.g., quartz, feldspar, mica, olivine, topaz, and garnet);

2) **sedimentary minerals:** those that resulted from the sedimentation processes in fresh or salt water (e.g., gypsum, salt, fluorite, and silvine);

3) **seam minerals:** those that were formed in veins in rocks. There are many types, and some were formed by thermal waters in which they were dissolved.

Iceland spar exhibits birefringence: When you look through it you see a double image.

Salt is a sedimentary mineral. The illustration shows a mountain of salt in Cardona, Spain.

PRECIOUS STONES

Certain hard, transparent minerals that are very valuable because of their rarity are known as precious stones; they are used in jewelry and in the decorative arts. There is usually a distinction between truly **precious** stones (diamonds, rubies, sapphires, etc.) and **semiprecious** ones (topaz, tourmaline, garnet, amethyst, etc.). Precious stones were considered luxury items in ancient times, and nowadays they fetch high prices in the marketplace.

amethyst

ruby

emerald

sapphire

QUARTZ

Quartz is a variety of silica crystallized in a hexagonal shape. It is one of the hardest minerals and, when cut properly, can be used as a piezoelectric oscillator: If subjected to an electric current, it vibrates in a very precise way and is used in making watches.

THE CULLINAN

The Cullinan diamond is the largest diamond in the world; it weighs 3,106 carats (621 g). It was found in 1905 in a mine close to Cullinan, Republic of South Africa and given to King Edward VII of Great Britain.

Minerals that are affected by electric current are known as **ferromagnetic**; an example is the magnetite used in compasses.

magnetite

Introduction

The origin
of the Earth

Geological
history

Crystallography

Minerals

Rocks

Activity of
the planet

Meteorology

Types of
climate

Seas and
oceans

Inland
waters

Landscape
formation

Erosion

Human
landscapes

Cartography

Subject index

THE CHEMICAL CLASSIFICATION OF MINERALS

The most precise classifications of minerals are those based on their crystallographic properties or their chemical composition. There are a great many systems. We will now take a look at some of the classifications that can be used.

(1) Native elements

Elements that occur in nature in their original state and have great economic value.

Examples include copper, silver, gold, lead, platinum, iron, diamond, graphite, and sulfur.

A gold ingot.

(2) Sulfides

Sulfides include sulfur compounds, selenium, arsenic, antimony, and tellurium. They are important in obtaining other minerals.

Examples include galena (lead), pyrite (iron), cobaltite (cobalt), antimonite (antimony), and argentite (silver).

Galena.

(3) Sulfosalts

Salts containing sulfur and another element (e.g., antimony, arsenic, or bismuth).

Examples include proustite, enargite, and bournonite.

Proustite.

(4) Oxides

Compounds consisting of oxygen combined with a metal. They commonly exist in amorphous masses and rarely as crystals. They are a convenient source of metals.

Examples are magnetite, crysoberyl, menite, corundum, sapphire, ruby, bauxite, and limonite.

Bauxite.

(5) Halides

Compounds consisting of fluorine, chromium, bromine, or iodine.

Examples are gem salt and common salt, silvine, carnalite, and fluorite.

Common salt.

(6) Carbonatos

Minerals composed of carbon and oxygen, along with other elements.

Examples are calcite, marble, magnesite, dolomite, and azurite.

Azurite.

(7) Nitrates

Compounds of nitrogen and oxygen with another element. Generally nitrates are water-soluble.

An example is Chilean saltpeter.

A poster advertising Chilean saltpeter, used as a fertilizer.

(8) Borates

Compounds of oxygen, boron, and metals.

Examples include borax, rasorite, and ulexite.

Borax is commonly used for decorating porcelain.

(9) Phosphates, vanadates, and arseniates

Compounds of phosphorus, vanadium, or arsenic.

Examples include apatite, turquoise, and camotite.

Turquoise.

(10) Sulfates

Compounds of sulfur, oxygen, and another element.

Examples include gypsum, anhydrate, baritine, epsomite, and alunite.

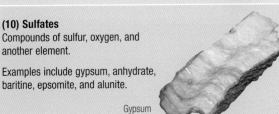
Gypsum

(11) Chromates, molybdenums, and wolframates

Compounds of chromium, molybdenum, or wolfram.

Examples are wolframite and crocoite.

Wolframite is used in making the filament in light bulbs.

(12) Silicates

Compounds involving silicon and other elements. They make up 95 percent of the Earth's crust.
Examples include olivine, cyanite, topaz, garnet, zircon, emerald, aquamarine, tourmaline, jadeite, amianthus, asbestos, and rhodonite.

Topaz.

(13) Radioactive minerals

Compounds that emit radiation.

Examples include uraninite, thorianite, branerite, carnotite, and thorite.

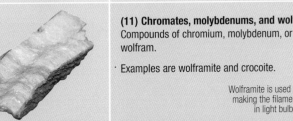
Pitchblende (a source of uranium).

ROCKS

Rocks are aggregates of various minerals; in some cases, though, they can be made up of a single mineral. They were formed in different ways at different depths, but they can be found everywhere in the Earth's crust because they appear in the form of outcrops. Rocks are divided into three large groups depending on how they were formed: igneous, metamorphic, and sedimentary.

IGNEOUS ROCKS

Igneous rocks are produced when the melted **magma** inside the Earth rises to higher levers where the temperature is lower and where it cools and solidifies. They are also produced when parts of the crust cave in and melt, subsequently returning to the surface in the form of a melted mass that then solidifies. When cooling occurs very quickly, there is no time for **crystal** formation, but if cooling takes place slowly, the longer the process, the better the crystals.

MAGMA

Magma is a mass of melted rock that is found at great depths. When it appears on the surface it is called lava.

PERIDOTITE

Peridotite is an extremely basic rock because it contains less than 45 percent silica.

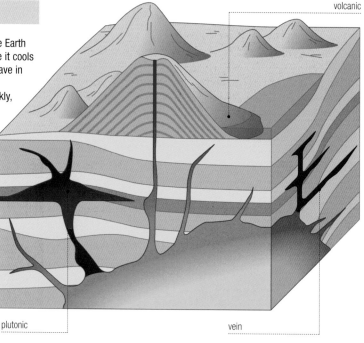

volcanic

plutonic

vein

Igneous rocks are produced by the solidification of melted magma that rises to the earth's upper layers.

TYPES OF IGNEOUS ROCKS

If we categorize them according to how they were formed, we have these possibilities:

1) **plutonic:** they crystallize inside the crust and form great masses of regular crystals;

2) **vein:** they solidify in the cracks through which the magma moves and form large and small crystals;

3) **volcanic:** they solidify on the outside after volcanic eruptions and form few crystals. They are classified according to the amount of silicon they contain:

a) **acid:** more than 66 percent silicon (e.g., granite);

b) **intermediate:** between 52 percent and 66 percent silicon (e.g., andesite);

c) **basic:** between 45 percent and 52 percent silica (e.g., basalt).

Many rocks are named for the place where they are found, such as hawaiite, which was first discovered in Hawaii.

Granite.

Basalt is a volcanic rock that sometimes forms spectacular deposits of hexagonal columns.

METAMORPHIC ROCKS

Metamorphic rock is of extremely varied composition and is produced by the transformation (**metamorphosis**) of some other type of rock. This occurs when a rocky mass is subjected to conditions of high pressure or temperature that cause its crystallographic structure to change (**recrystallization**), giving rise to new minerals. Many metamorphic rocks exhibit **foliation** by breaking along plat planes. Examples include marble, quartzite, slate, gneiss, and amphibolite.

FOLIATION

Foliation involves the alignment of minerals along a plane.

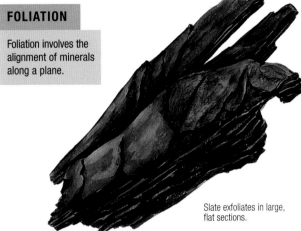

Speckled schist, an example of a metamorphic rock.

Slate exfoliates in large, flat sections.

CLAST

A clast is a particle or piece of mineral trapped in sediment.

CLASSIFICATION OF METAMORPHIC ROCKS BY TEXTURE

Type of Grain	Type of Rock	Type of Foliation
fine	slate	almost flat and very compressed
fine	phylite	undulating
medium	schist	undulating planes
coarse	gneiss	slightly separated planes
coarse	magmatite	not well developed

CLASSIFICATION OF SEDIMENTARY ROCKS BY GRAIN SIZE

Size of Grain (Inches)	Name of Clast	Type of Rock
>10	block	conglomerate
0.08–10	pebble	conglomerate
0.0025–0.08	sand	sandstone
0.00008–0.0024	mud	limolite or arcilite
<0.0008	clay	limolite or arcilite

A hunk of nummulitic limestone containing encrustations of nummulites.

SEDIMENTARY ROCKS

Sedimentary rocks are formed from minerals that accumulate in a particular place because of erosion. Afterward, the sediment becomes compacted, clings together, and changes chemically. There are several types:

1) **detrital:** formed by residues (e.g., arsenic, limolite, arcilite);

2) **chemical in origin:** formed by the precipitation of mineral elements from water (such as tufa, travertine, and evaporite);

3) **marl:** mixtures of detrital rocks and rocks of chemical origin;

4) **organogenetic:** produced by the accumulation of skeletonized, calcareous residues, generally from plankton (e.g., chalk, diatomite, and nummulitic limestone).

Tufa is a mineral of chemical origin.

Introduction

The origin of the Earth

Geological history

Crystallography

Minerals

Rocks

Activity of the planet

Meteorology

Types of climate

Seas and oceans

Inland waters

Landscape formation

Erosion

Human landscapes

Cartography

Subject index

VOLCANISM

One of the most notable manifestations of the planet's activity is the eruption of **volcanoes**. There are different types of volcanoes, depending on how the lava is expelled, and they are located in only certain parts of the globe. Volcanoes are also the only place where we can gain access to materials from the interior of the crust or the mantle, and so they are of great interest to science.

THE FORMATION OF VOLCANOES

Volcanoes are the exit points for molten rocky material from the inside of the Earth. Sometimes this material comes from the depths of the **crust**, where it was formed as a result of high pressure and temperature. On other occasions, the melted mass comes straight from the **mantle**. Volcanoes are not located in all areas but only in those where there is more fusion activity due to a **crust plate** that is being destroyed. This can happen in such places as the Pacific coastlines, where the crust is sliding underneath the continents. In the Atlantic, on the other hand, this activity is much less pronounced because new crust is being formed.

Cotopaxi volcano (19,342 ft [5,897 m]) in the Andes of Ecuador.

SOME OF THE EARTH'S MAJOR VOLCANOES

Volcano	Altitude (feet)	Location
Cotopaxi	19,342	Ecuador
Erebus	12,444	Antartica
Erta Ale*	9,699	Ehtiopia
Etna*	10,808	Italy
Fuego	12,343	Guatemala
Fujiyama	12,385	Japan
Kilauea	4,038	Hawaii
Kilimanjaro	19,335	Tanzania
Krakatoa (new)*	2,729	Indonesia
Lasen Peak	10,450	California
Llullaillaco	22,100	Chile
Mauna Loa	13,638	Hawaii
Nevado del Ruiz	17,712	Colombia
Niragongo	11,378	Zaire
Ojos del Salado*	22,609	Chile-Argentina
Orizaba	18,696	Mexico
Popocatepetl	17,883	Mexico
Stromboli*	3,037	Italy
Teide	12,195	Canary Islands
Vesuvius	4,165	Italy
Villarrica	9,315	Chile

*Still active.

→ When molten material reaches the Earth's surface, it is called lava.

TYPES OF VOLCANOES

Lava doesn't always come out in the same way, and it can take different shapes. That's one way of distinguishing among types of volcanoes.

Hawaiian: forming a broad, shallow cone; the lava is very liquid, and the eruption is not very violent.

Strombolian: forming a more or less regular cone, with a single exit shaft; the eruption is very violent.

Vulcanian: forming a fairly regular cone, but the outlet has smaller branches; the eruption is very violent.

Pelean: The lava is very dense, and it forms a solid column as it reaches the surface.

LAVA

Lava is made up of the melted material in the mantle and the parts of the crust that melt in certain parts of the globe. Lava also contains many gases such as **water vapor** (which commonly makes up a little more than half), **carbon dioxide**, **hydrogen sulfide**, **hydrochloric acid**, **hydrogen**, and **carbon monoxide**. All these gases are dissolved in the magma because of the tremendous pressure, but when they reach the surface, they escape just like the bubbles of a carbonated drink when the bottle is opened. When the lava rises quickly, the gas escapes with great force and the eruption is very violent; when the lava rises slowly, the gas escapes little by little.

SOME OF THE MAIN VOLCANIC ERUPTIONS THAT HAVE TAKEN PLACE AROUND THE WORLD

Year	Volcano	Deaths
79 B.C.	Vesuvius (Italy)	4,000-5,000
1586	Kelut (Indonesia)	10,000
1621	Vesuvius (Italy)	5,000
1783	Laki (Iceland)	9,350
1792	Unzen (Japan)	14,300
1815	Tambora (Indonesia)	95,000
1883	Krakatoa (Indonesia)	36,400
1902	Mt. Pelée (Martinique)	29,000
1980	Mt. St. Helens (United States)	57
1982	El Chichon (Mexico)	1,900
1985	Nevado del Ruiz (Colombia)	23,000
1986	Lago Nyos (Cameroon)	2,000
1991	Pinatubo (Philippines)	800

PARTS OF A VOLCANO

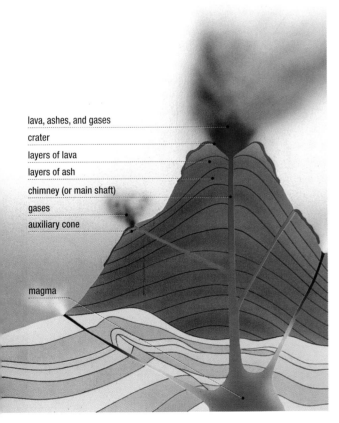

lava, ashes, and gases
crater
layers of lava
layers of ash
chimney (or main shaft)
gases
auxiliary cone

magma

The lava flows down the sides of the volcano and forms rivers of molten material.

On May 27, 1883, the eruption of the volcano Parbuatan (which was heard 3,000 miles [5,000 km] away) caused a large part of the island of Krakatoa to disappear; the island was split into several pieces, and more than 36,000 people were killed.

 The word *volcano* comes from Vulcanus, the name of the ancient Roman god of fire and metallurgy.

EARTHQUAKES

Even though earthquakes are less spectacular than volcanic eruptions, they are another manifestation that we can perceive of the activity that takes place inside the earth. In addition, the damage they cause can be much more severe and produce more victims because they often are unpredictable. The areas that have the greatest number of earthquakes are the same ones that have the greatest amount of volcanic activity.

WHAT CAUSES THEM?

The **plates** that make up the earth's crust are under tension. The area where two of these plates come together is called a **fault**. Each plate moves independently of the other along the fault line, but the grinding that takes place sometimes makes them stick together. When the pressure becomes very high, the force exceeds the holding capacity of the plates and both of them suddenly move, making the ground shake and releasing an enormous amount of energy. This entire phenomenon is known as **seismic activity** or an **earthquake**.

→ The science devoted to the study of earthquakes is called **seismology**, and the scientists in the field are **seismologists**.

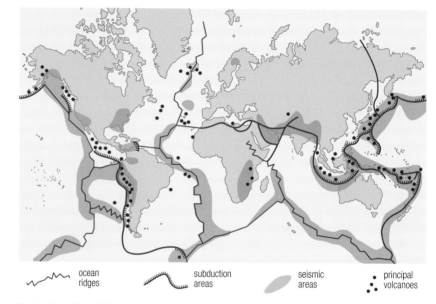

| | ocean ridges | | subduction areas | | seismic areas | | principal volcanoes |

The locations of the earth's main volcanoes, which coincide with the oceanic ridges and with seismic areas.

MEASURING AND PREVENTING EARTHQUAKES

Throughout the centuries efforts have been made to predict earthquakes. This work involves studying the slight movements produced in the earth's crust to determine when they may become strong. The device used is called a **seismograph**, and thousands of these instruments are installed all around the globe. In countries where earthquakes are common, such as **Japan**, construction is traditionally light in order to minimize earthquake damage. Modern buildings are constructed with special frames that allow them to vibrate without collapsing. The **skyscrapers** in Tokyo have a type of spring built into their foundations to help absorb shocks.

FOCUS AND EPICENTER

The **focus** of an earthquake is the area where a fault ruptures and the earthquake begins; the **epicenter** is the area on the surface located precisely above the earthquake's focus.

SOME SIGNIFICANT EARTHQUAKES

Year	Location	No. of Victims
1906	San Francisco, United States	500
1923	Tokyo, Japan	143,000
1976	Tangshan, China	250,000
1980	Avellino, Italy	8,000
1985	Mexico City, Mexico	30,000
1990	Northern Iran	40,000
1995	Kobe, Japan	5,000
2000	Western Turkey	30,000

THE PARTS OF A SEISMOGRAPH

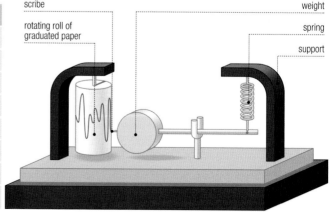

scribe

weight

rotating roll of graduated paper

spring

support

Introduction

The origin
of the Earth

Geological
history

Crystallography

Minerals

Rocks

**Activity of
the planet**

Meteorology

Types of
climate

Seas and
oceans

Inland
waters

Landscape
formation

Erosion

Human
landscapes

Cartography

Subject index

SEAQUAKES AND TIDAL WAVES

When the **epicenter** of an earthquake is located at the bottom of the **ocean**, it causes movements in the water that are perceptible at the surface. The waves that head toward the closest coastline are a little more than 3 ft (1 m) high. However, when the waves reach shallow waters, they start to grow larger (33 ft [10 m] or more), and when they slam into the shore, they cause tremendous damage. These huge tidal waves are called **tsunamis**. Underwater volcanic eruptions also cause catastrophic tsunamis.

The tsunami caused by the explosion of Krakatoa reached a height of 115 ft [35 m].

Tsunamis are common in Hawaii. The largest recorded to date was 56 ft (17 m) high.

In large cities located in seismic areas, skyscrapers are constructed with a frame that resists temblors.

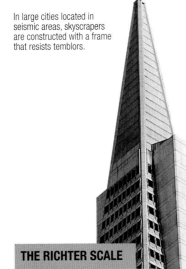

THE RICHTER SCALE

The Richter scale classifies earthquakes by number according to the amplitude of the waves and the amount of energy released.

Damage caused by an earthquake.

IN CASE OF AN EARTHQUAKE

If you find yourself in a building at the time an earthquake occurs, there are some measures you can take to avoid injury:

1) Take cover on the floor or under a table to protect yourself from objects that might fall from the ceiling.

2) Move away from windows and large objects that may fall.

3) Do not use the elevator.

4) Do not go down the stairs while the temblor is in progress.

5) Try to turn off any kind of device that might cause a fire. If you are outdoors, move away from electric wires and buildings because debris and windowpanes may fall.

THE MERCALI SCALE
(CLASSIFIES EARTHQUAKES ACCORDING TO THE DAMAGE THEY PRODUCE)

Intensity	Damage Produced
I	Detectable only by some seismographs.
II	Detectable only by some people in tall buildings.
III	Hanging objects move, and many people notice the quake.
IV	Some walls creak, and tables move; people indoors notice it, but it is scarcely noticeable outdoors.
V	Some windows break, hanging objects fall down, and nearly everyone notices it.
VI	Moderate damage occurs, with fallen objects, displaced furniture, and loosening of false ceilings.
VII	The walls of some buildings collapse, chimneys fall; buildings constructed specifically to resist earthquakes do not fall, but it is difficult for people to remain standing.
VIII	Slight damage occurs in specially constructed buildings, and partial collapse in others; chimneys, walls, and monuments topple.
IX	Buildings of normal construction collapse, specially constructed buildings experience significant damage, and buildings separate from their foundations.
X	Serious damage and collapse of specially constructed buildings occurs; others are reduced to rubble, and railroad tracks are twisted.
XI	All buildings collapse, a few fragments are left standing, and railroad tracks are badly twisted.
XII	Total destruction, changes in the landscape, objects thrown far away.

SEDIMENTS AND STRATIFICATION

The result of erosion on the Earth's crust is fairly large deposits of rocky residue, which accumulate in low places because of gravity, wind, or the action of water. This activity produces sediments, which become arranged in layers and eventually turn into new sedimentary rocks.

The continuous passage of water through a certain type of terrain eventually cuts deep canyons; the materials that are washed away are deposited in the lower reaches of the river.

SEDIMENTATION

Depending on the force of **erosion**, the surface of the Earth's crust breaks up into pieces that can be large **blocks** of rock, coarse **gravel**, or very fine **silt** with granules less than 0.039 in. (1 mm) in diameter. The force of gravity causes this material to collect in low places, sometimes in large quantities, building up successive layers known as **strata**. The type of stratum depends on the type of erosion at any given time (water, wind, etc.). The depth of the layer depends on the amount of time over which the deposits occur, but it can exceed several miles (kilometers).

During the Carboniferous age organic remains settled as sediments; they produced the petroleum and coal that we consume today.

STRATIGRAPHY

The branch of geology that focuses on the sedimentary strata or the layers of the Earth that make up the crust.

Marshes are swampy areas of brackish water located near the mouth of a river where it empties into the ocean. They generally contain silt and clay from higher areas.

FROM SEDIMENT TO ROCK

The first layer of sediment (the superficial stratum) is commonly soft; however, as new layers are deposited on top, the weight increases and so does the pressure exerted on the sedimentary particles. This causes the mass to settle slowly at the same time that it becomes compressed. This physical pressure, combined with an increase in **temperature**, eventually causes chemical changes, and the soft sediment turns into **hard rock**. However, the temperature is low enough so that the material isn't completely transformed; in that case it would melt and become **metamorphic rock**.

Underground deposits of water (aquifers) form primarily in sedimentary soils.

Strata often are found pitched at an angle, which means that they were subjected to movement after their formation.

WHERE DOES SEDIMENTATION OCCUR?

Sedimentary processes can occur in any place on the Earth's surface where erosion occurs, but not all the material deposited ends up becoming sedimentary rock, because if erosion continues in the same place, it can also sweep away the sediments before they fuse. There are three main areas where sedimentary processes occur:

1) **marine:** deposits form on the continental platform and in deep-sea areas;

2) **continental:** deposits form at the foot of mountain chains, in glaciers, along rivers, and in deserts;

3) **transitional:** sedimentation occurs at contact points between sea and land, as in deltas and marshes.

Organic remains must be deposited in sediments in order to become fossilized and are preserved only if the sediment changes to sedimentary rock.

The sediments carried down by a river are deposited in its lower reaches, forming flat ground in the places where the river cuts through to form meanders.

BIOSTRATIGRAPHIC UNITY

A series of strata that contain a characteristic group of fossils demonstrate biostratigraphic unity.

A cross section of a rock that contains fossils allows geologists to study antiquity and the formation of the terrain.

SEDIMENTS AS INDICATORS OF TIME

The thickness of the sediment makes it possible to deduce how long it took to form if the speed of sedimentation can be determined indirectly. In addition, each type of sediment is characteristic of an epoch (of rains, desert, glaciers, etc.). All this information helps geologists to determine the age of a thick sedimentary layer and therefore the age of the fossils it contains. If the sediment is not altered by changes (slides, faults, etc.), the upper strata are always the most recent ones.

Deltas are formed at the mouth of a river by the accumulation of materials that the river carries downstream. In general, deltas are very fertile ground that is cultivated intensively.

FOLDS

The Earth's crust is solid, but because new parts of it are continually being formed and destroyed, there are tremendous forces inside it that change its shape. These forces work over the course of millions of years and cause waves in the crust, which are known as folds. When the acting forces are so powerful that the material's elasticity cannot withstand them, the folds rupture.

THE FORMATION OF FOLDS

The rocky materials that make up the Earth's crust have a certain degree of elasticity; it is greatest in soft **sedimentary** rocks and least in **metamorphic** rocks. When intense forces come into play, for example, when two continents collide, the rock yields and flexes, folding into a curved shape corresponding to the force that is exerted. These **folding** processes can occur at shallow depths in the crust, and they are the cause of the globe's large **mountain chains**. Depending on the elasticity of the rock, the fold bends as far as it can, but when its folding capacity is exceeded, it breaks and produces a **fault**.

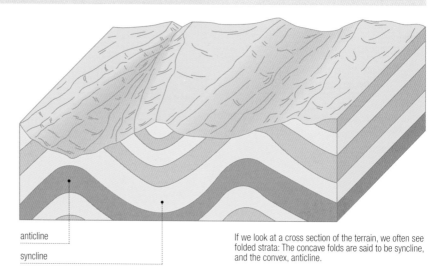

anticline

syncline

If we look at a cross section of the terrain, we often see folded strata: The concave folds are said to be syncline, and the convex, anticline.

 The Appalachians were formed when the earth's crust buckled from the collision of North America and Africa about 300 million years ago.

ELASTIC DEFORMATION

When the deforming force recedes, the rock recovers its original shape and no fold is formed.

A clear example of an anticline fold.

THE PARTS OF A FOLD

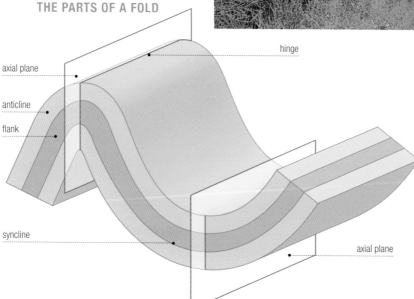

axial plane

anticline

flank

hinge

syncline

axial plane

THE PARTS OF A FOLD

In the simplest cases, when a fold forms a simple curve, we distinguish the point at which the direction of the curve changes (the point of inflection) and the areas on both sides of it. If the fold is convex at the point of inflection and is shaped like an arch, it is said to be **anticline**; if it is concave and U-shaped, it is called **syncline**. The point of inflection is called the **hinge**, and its two sides are the **flanks**. The surface that passes through the area of greatest curvature is called the **axial plane**, and the line that cuts between the axial plane and the hinge is the **axis**.

TYPES OF SIMPLE FOLDS

When we look at a cross section of terrain we can see all kinds of folds. Some can scarcely be detected, because the forces that formed them were very weak or just beginning to act. In other instances, though, the terrain is folded numerous times and at acute angles. A **monocline fold** has just one gentle curve. An **anticline fold** is convex at the top; a **syncline fold**, on the other hand, is concave at the top. A **dome** is an anticline that has nearly symmetric flanks.

Many times road construction uncovers folds.

Rocks often exhibit microfolds.

FLEXION AND TERRACING

A **flexion** is a deformation in the terrain that stops short of forming a fold; a **terrace** is an area of slight incline (largely level) in terrain that has a consistent pitch.

ISOCLINES

Isoclines are sets of folds arranged in a regular pattern at the same angle and following the same direction.

COMPLEX FOLDS

When the compression forces that caused the fold are very strong but do not cause a break in the terrain, the result is complex folds. The most basic ones are a succession of **anticlines** and **synclines**, which give the countryside an even, undulating appearance. In other places anticlines and synclines follow in succession but have different shapes. This may result in anticlines with a very pronounced curvature and more gentle synclines, or the converse. In other instances, they become superimposed on one another and are called **overturned folds**. When the forces are particularly strong, some folds rupture and form mixed structures involving both **folds** and **faults**.

Often, folds are subtle and as broad as they are long.

Formation of the different types of folds: (1) Uniform pressure produces a simple fold, or a symmetric anticline. (2) Unequal pressure can form an asymmetric anticline. (3) Continuous pressure on an asymmetric anticline can produce a vaulted fold. (4) A vaulted fold can rupture and fall in on itself; in such cases it is called an overturned fold. (5) When an overturned fold is displaced significantly, the result is a stratum.

dome

basin

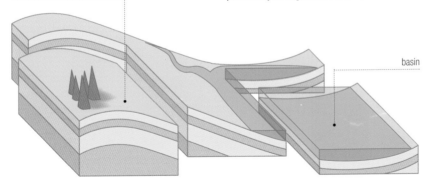

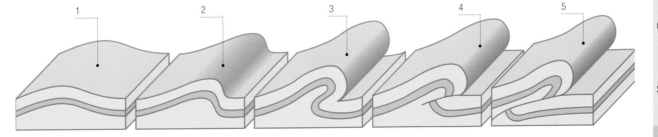

1   2   3   4   5

FAULTS

One of the easiest features of the terrain to observe is faults, or ruptures in a fold, especially if the terrain is of a sedimentary nature. This type of deformation ends in a rupture and produces a great number of geological features. In addition, in many instances when the sides are displaced suddenly, there is an earthquake.

WHEN THE GROUND BREAKS UP

The materials that make up the Earth's crust are flexible to a certain extent. When they are subjected to force they may become distorted and produce a **fold**, but sometimes the force exceeds the deformation capacity of the ground, and instead of folding it breaks. This produces what is called a **fault**. The part of the crust on which the force acts then moves over the part that resists it along the **fracture plane**. If the terrain is sedimentary, we can see that the strata separate along a line. This line is the plane along which the fault splits apart.

THE PARTS OF A FAULT

The fracture plane that produces a fault is called the **fault plane**. It is visible on the surface of the earth as a fairly straight line, which is known as the **fault line**. The fault plane can be nearly perpendicular or at a pronounced angle. When it is perpendicular, the two sides of the fault slide along one another, but when the fault plane is on a slant, one of the masses of rock slides over the other. The block located above the fault plane is known as the **roof**, and the one below it is the **wall**. These are the four essential features of a fault.

The San Andreas fault is a spectacular fracture more than 248 miles (400 km) long that extends from the Gulf of California to north of the city of San Francisco.

THE PARTS OF A FAULT

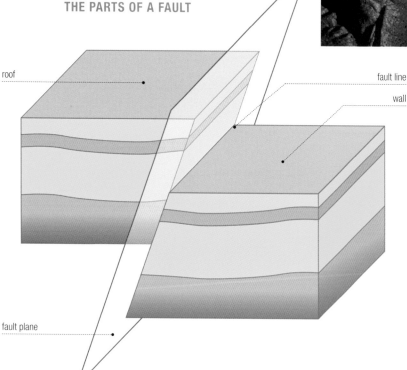

roof

fault line

wall

fault plane

A fault can be caused by a quick, intense force or by a small force that acts over a long time.

There are faults that are hundreds or even thousands of miles (kilometers) long.

TYPES OF FAULTS

Because the blocks that make up the Earth's crust are not uniform but contain different types of rocks of varying resilience, the response to a force may be quite different in various locations. However, to understand faults more thoroughly, we can classify them as several basic types. A **normal fault** is one in which the roof is lower than the wall (because it slides along the fault plane). In an **inverse fault**, the roof is higher than the wall (because compression forces push it upward along the fault plane). In a **tear fault**, both rocky masses move laterally.

FAULT MIRROR

The area of a fault plane that becomes polished by the grinding of the two moving masses of rock is called a fault mirror.

FAULTS DO NOT OCCUR IN ISOLATION

When great forces act on the terrain, often there are areas that fold and others that rupture. As a result there is a very characteristic relationship between **folds** and **faults**. When several faults occur in a single area, all of them together make up a **fracture zone**. Many times there are successions of parallel faults, or faults arranged in a radial pattern or even in a type of grid. In other instances, the faults are arranged in steps. At the bottom of valleys produced by faults it is common to find a series of long lakes, as in eastern Africa (e.g., Malawi, Tanganyika, Rudolph, and other lakes).

SOME TYPES OF FAULTS

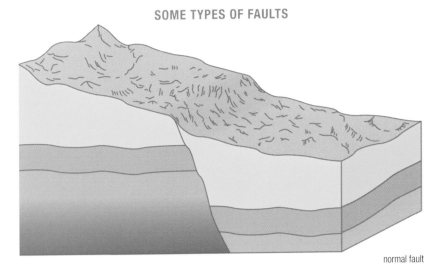

normal fault

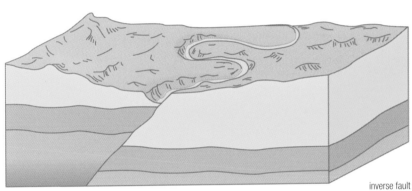

inverse fault

tear fault

The Rift Valley is made up of a series of depressions and cuts in the ground (in the photo, near Lake Natron in Tanzania) that extend from Jordan to the lower third of the African continent.

 In North America, California is moving toward the northeast along the San Andreas fault. As a result, this is an area of considerable seismic activity.

THE FORMATION OF MOUNTAINS

Even though mountains appear huge to us, when they are viewed from outer space, or even from an airplane, they appear to be a small alteration in the relief of the Earth's crust. However, they are extremely important for the life of all organisms. Mountains are the result of the Earth's crust folding and rising in elevation in certain areas.

THE FORMATION OF A MOUNTAIN CHAIN

There are many theories concerning the formation of mountains. The most recent ones hold that they are the result of a collision between the great **plates** of the Earth's crust. It all starts when two of these plates separate, leaving a low spot between them. Then, so much **sediment** collects in the hollow that the whole area caves in, causing both plates to be pulled together. As they come closer to one another, the layer of sediment folds, and buckles in an arch. Finally, to balance the various masses of the crust, the **fold** rises and forms a mountain chain. This can also happen when the plates collide at their edges, without a preexisting fold, causing the crust to buckle upward and form **collision chains**.

Orogeny is the branch of geology that deals with mountain formation.

In general, old mountains look smoother than more recent ones.

THE AGE OF MOUNTAINS

When you travel, you encounter many types of mountains. Some have a high, pointed **peak**, and others have a rounded or flat **summit**. The former are more recent because they have not yet been subjected to erosion, which removes material over the course of millions of years and smoothes out shapes as if they had been "filed down," which is what has happened to old mountains.

Hills are small, isolated elevations.

THE OROGENIC CYCLE

Each of the major periods in the formation of a mountain chain is known as an orogenic cycle.

OROGENIC CYCLES

Throughout the long geological history of the Earth there have been four great **orogenic cycles**, and they produced all the mountains that now exist. There are very few remains from the oldest cycles, because **erosion** and new **folds** produced new mountains. The last of these cycles is known as the **alpine orogeny** because that is when the Alps in Europe, the Andes in South America, and the Rocky Mountains in North America were formed.

Spain's highest mountain is Teide, in the Canary Islands; it is 12,195 ft (3,718 m) high. On the Iberian Peninsula, the highest mountain is Mulhacen (shown in the photo), in the Sierra Nevada, at 11,408 ft (3,478 m).

Introduction

The origin
of the Earth

Geological
history

Crystallography

Minerals

Rocks

**Activity of
the planet**

Meteorology

Types of
climate

Seas and
oceans

Inland
waters

Landscape
formation

Erosion

Human
landscapes

Cartography

Subject index

SOME OF THE EARTH'S HIGHEST MOUNTAINS*

Continent	Mountain	Altitude (ft)	Mountain Chain	Location
Europe	Mont Blanc	15,767	Alps	France
	Monte Rosa	15,213	Alps	Italy, Switzerland
	Breithorn	13,661	Alps	Italy, Switzerland
	Jungfrau	13,638	Alps	Switzerland
	Aneto	11,165	Pyrenees	Spain
	Monte Perdido	11,004	Pyrenees	Spain
	Etna	10,808	Etna	Sicily (Italy)
Asia	Everest	29,015	Himalayas	China, Nepal
	K2	28,244	Karakorum	India
	Kanchenjunga	28,201	Himalayas	India, Nepal
	Lhotse	27,916	Himalayas	Nepal
	Makalu	27,814	Himalayas	China, Nepal
	Dhaulagiri	26,804	Himalayas	Nepal
	Nanga Parbat	26,653	Himalayas	India
	Annapurna	26,496	Himalayas	Nepal
	Pico Comunismo	22,288	Pamir	Tadjekistan
	Demavend	18,598	Elburz	Iran
	Elbrus	18,476	Caucasus	Georgia
	Ararat	16,941	Armenia	Turkey
	Fuji-Yama	12,385	Fuji-Yama	Japan
Africa	Kilimanjaro	19,336	Kilimanjaro	Tanzania
	Kenia	17,036	Kenia	Kenya
	Ruwenzori	16,790	Ruwenzori	Uganda
	Ras Dashan	15,154	Ethiopian Massif	Ethiopia
	Tubkal	13,655	Atlas	Morocco
	Cameroon	13,350	Cameroon Massif	Cameroon
The Americas	Aconcagua	22,826	Andes	Argentina
	Ojos del Salado	22,610	Andes	Chile-Argentina
	Illimani	22,573	Andes	Bolivia
	Tupungato	22,304	Andes	Argentina-Chile
	Pissis	22,235	Andes	Argentina
	Mercedario	22,206	Andes	Argentina
	Huascarán	22,104	Andes	Peru
	Llullaillaco	22,104	Andes	Argentina-Chile
	Coropuna	21,697	Andes	Peru
	Chimborazo	20,425	Andes	Ecuador
	McKinley	19,844	Alaska	United States
	Cotopaxi	19,342	Andes	Ecuador
Oceania	Mauna Kea	13,802	Mauna Kea	Hawaii
	Cook	12,346	New Zealand Alps	New Zealand
	Kosciusko	7,308	Snowy	Australia
Antarctica	Erebus	12,444	Erebus	Antarctica

*This table contains only a selection of the most representative
mountains of the various continents and mountain ranges.

Climbing the highest mountains and scaling their most
difficult faces is an athletic challenge reserved only for
those mountaineers and climbers who are the best
prepared technically, physically, and psychologically.

The 40 highest mountains
in the world are in the
Himalayan chain and
the surrounding area.

The granite Paine massif (in the south of Chile, on the
border with Argentina) is the end of the Andes mountain
chain. Even though it does not exceed 9,840 ft (3,000
m), it consists of abrupt mountains, with glacial cirques
and valleys.

CONTINENTAL DRIFT

So far we have seen that when forces act on the Earth's crust, it folds up and produces wrinkles or ruptures and forms faults. We have also seen how these movements in the crust produce high mountains. All of this activity is due to a process discovered at the start of the twentieth century: the movement or drift of the continents, which is based on plate tectonics.

AN INCREDIBLE THEORY

In 1912 when **Alfred Wegener** put forth his theory concerning the origin of the present continents, most scientists criticized him openly and public opinion was not very positive. Wegener held that the **continents** are "floating" on a fairly dense portion of the **upper mantle**, which allows them to move. Based on paleontological evidence and the coincidence of geological material in separate continents (e.g., Africa and South America), he asserted that the present-day continents are the result of an original continent that broke up and whose fragments are moving around.

ALFRED WEGENER

A German geophysicist and explorer, Alfred Wegener lived from 1880 to 1930. He died in Greenland during his last expedition to the polar regions.

DIAGRAM OF PLATE MOVEMENT

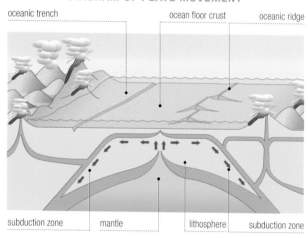

oceanic trench · ocean floor crust · oceanic ridge

subduction zone · mantle · lithosphere · subduction zone

PLATE TECTONICS

When **Wegener** put forth his theory, the depths of the oceans had not yet been explored, and although he was convinced that the continents were drifting apart, he was unable to explain why. In the mid-1970s, a theory known as **plate tectonics** was developed to explain this movement. It asserts that the earth's crust is made up of a series of rigid plates that move on the upper layer of the **mantle**. Furthermore, all the plates mate up in such a way that there are no low spots that allow direct contact with the mantle. These movements are the cause of the earthquakes that we detect on the surface.

SUBDUCTION ZONE

An area where one plate slips under another one is called a subduction zone.

When two plates meet in a subduction zone, one of them ends up being destroyed.

200 million years ago

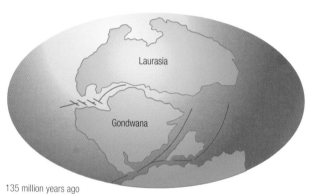

135 million years ago

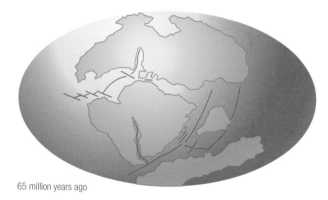

65 million years ago

Presently

THE OCEANIC RIDGES

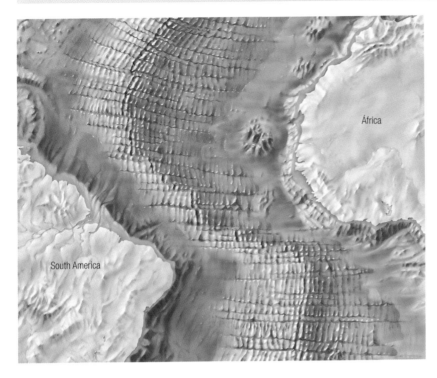

When the bottom of the oceans was explored, scientists were surprised to discover considerable variation in the terrain. In the middle of the oceans there is a long chain of mountains extending for several thousand miles (kilometers), and as high as 19,680 ft (6,000 m); this feature is known as the **oceanic ridge**. The mountains contain a lot of very intense **volcanic activity**, and geological studies have shown that they are precisely the result of the formation of new ocean floor. The mountain ranges of all the oceans follow a path that is more or less equidistant from the adjacent major continental masses, and the farther away from them, the older the rocks on the bottom.

PLATE COLLISIONS

When the **plates** come into contact with one another, the forces that move them work against each other and produce a collision; the result is that one of the plates slides beneath the other. The geological makeup of the **oceanic** and **continental** plates is different. The oceanic plates are denser than the continental plates. As a result, when two plates collide, it is always the oceanic that slips under the continental in a process known as **subduction**. When two oceanic plates collide, one still moves under the other, but this struggle can be won by either of the plates.

A map showing the floor of the Atlantic, with the mid-Atlantic ridge in the center

A QUESTION OF ANTIQUITY

The oceanic crust is continually being recycled. The oldest known material is just 190 million years old. The oldest continental crust is more than 4 billion years old.

The crust is divided into seven main plates and six or seven smaller ones. ←

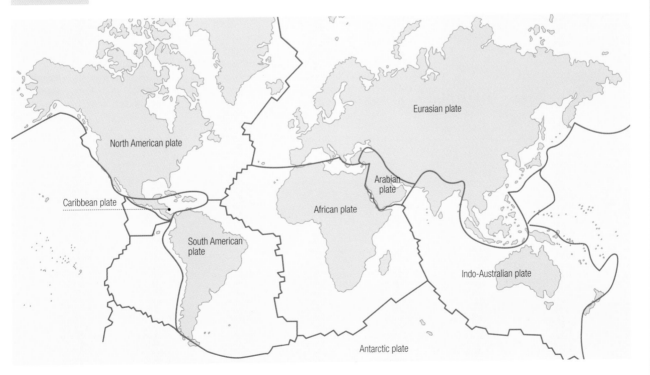

41

WEATHER AND CLIMATE

In contrast to what happens with geological phenomena, which take place so slowly that we can hardly perceive any difference, the atmosphere is in a state of continual change, with some changes occurring within a matter of minutes. These changes affect our well-being and health directly, and there are two branches of science devoted to them: meteorology and climatology, which take different approaches in dealing with the same material.

METEOROLOGY

Meteorology is the science that deals with phenomena that occur over the short term in the lower layers of the **atmosphere**, in other words, where plant and animal life occurs. This means that meteorology studies the atmospheric changes produced at any time according to such parameters as air **temperature**, **moisture content** of the air, **atmospheric pressure**, **wind**, and **precipitation**. The purpose of meteorology is to predict what the weather will be like over the next 24 or 48 hours and to make medium-range **predictions**.

The meteorological conditions on any specific day can be very different from the region's climate.

A weather vane (left), often in the shape of an animal or an arrow, indicates the direction of the wind. An anemometer (right) indicates the wind speed.

DYNAMIC METEOROLOGY

The study of the laws that govern the movements of the large air masses in the atmosphere is called dynamic meteorology.

Time and weather tell us the condition of the atmosphere at any given time and place.

THE STUDY OF WEATHER AND ITS APPLICATIONS

Meteorology studies the atmosphere from three different viewpoints. On the one hand, it describes the general conditions that exist in all areas of the globe, indicating the existence or absence of precipitation, the temperature, the pressure, and so forth. Second, it investigates how the large **air masses** behave in order to determine the **laws** that govern them. Thanks to these laws, it is possible to predict the weather. Meteorology has several practical applications; some of the most important ones are **aeronautical meteorology** (for regulating air traffic), **agricultural meteorology** (for predicting the best conditions for planting and harvesting), and **medical meteorology** (which deals with the effect of these factors on human health).

Temperature distribution around the globe: always cold (in blue); temperate summers and cold winters (purple); hot summers and cold winters (pink); cool summers and mild winters (green); hot summers and temperate winters (yellow); always hot (orange).

CLIMATOLOGY

The science of climatology studies the **climate** and its variations through time. It resembles meteorology because it studies the same factors (precipitation, humidity, wind, pressure, etc.), but it is different because it focuses on the **long term**. It defines climate as the series of meteorological phenomena that characterize the usual conditions of a certain place on the surface of the globe. Climate represents a **statistical value**—it indicates the most common conditions and therefore the ones that we are most likely to encounter over a long period of time. Thus, a desert may experience a storm with a large amount of water, but only every few years. Despite this regular rain, the climate is still a desert one.

The focus of climatology, the study of the climate, allows scientists to observe weather changes in specific areas. In recent decades, for example, they have discovered that deserts are increasingly getting less rain and that temperatures are rising in polar regions.

CLIMATIC CYCLE

A climatic cycle is a long period of time (many years or centuries) over which climatic conditions take place.

Meteorological satellites make it possible to observe the atmosphere and its variations, contributing to climate studies and weather prediction.

MAPS

The geographical maps we have today are the result of centuries of experience. At first maps were made by noting the locations where travelers on land or sea encountered **geographical features** (capes, bays, islands, mountains, valleys, etc.). They were later improved through the use of more precise measuring devices. Nowadays maps are made with the aid of **photographs** taken from outer space by **satellites**. Maps are an essential element in studying weather and climate because variations in atmospheric conditions depend in large measure on the location of seas and continents and on the existence of geographical features.

The weather varies from one day to the next. Climate varies over long periods of time—several centuries.

longitude 0

latitude 0 (equator)

Map of the Earth (Mercator projection). The vertical lines represent meridian lines or longitude, and the horizontal lines represent parallels, which are used in determining latitude.

THE CIRCULATION OF THE ATMOSPHERE

The atmosphere is a mixture of gases that surround the planet; they are very light, and it is easy for them to move around the surface. Like all gases, the atmosphere changes weight according to temperature, which also can make it rise to a certain degree.

Because there are temperature variations from one place to another on the Earth, the atmosphere heats up differently, and the result is that it is in constant motion.

The amount of heat that reaches the ground also depends on the transparency of the air.

 In the Northern Hemisphere the exposed land is the dominant feature; in the Southern Hemisphere, oceans predominate.

HEAT

The **energy** from the sun that reaches the Earth passes through the atmosphere before it warms the Earth's surface and is warmed in turn. But the Earth's surface may be water (such as the **oceans**) or rock (such as the **continents**), and water tends to warm up and cool down more slowly than rock. As a result, the continents cool off and heat up before the seas, and this is why different temperatures occur from one place to another.

The amount of energy that reaches the Earth also depends on the **angle** of the sun's rays. If they come down vertically (e.g., at noon or in the summer), they supply more heat than if they come down at an angle (e.g., at dusk or in the winter). In polar regions the sun's rays come down at a much greater angle than in equatorial areas. As a result, more heat accumulates in the latter than at the poles.

LATITUDE AND ALTITUDE

Latitude indicates the location of any point on the Earth's surface with respect to the equator and the poles. This distance is measured in **parallels**, that is, each of the smaller circles into which the globe can be cut horizontally. Parallels are measured in degrees, from 90 degrees for the pole to 0 degrees for the equator; these measurements serve as an approximate reference to the angle of the sun's rays. As a result, the **high latitudes** (from 60 degrees to 90 degrees) receive less solar energy than the low ones (from 0 degrees to 30 degrees).

Altitude refers to the height above sea level. As altitude increases, the density of the atmosphere decreases and so does its ability to absorb heat. As a result, the overall temperature decreases with respect to elevation, and so on any given day it is warmer at the foot of a mountain than at the summit.

Mountain peaks are commonly colder than valleys.

GENERAL CIRCULATION OF THE ATMOSPHERE

We have already seen that the surface of the Earth heats up differently depending on whether it is water or solid land, and that another factor is the amount of energy that arrives to provide heat, according to latitude. In addition, depending on **altitude**, the amount of heat absorbed is different and there is also a difference in **atmospheric pressure**. All this means that there are areas where lighter (warmer) air masses tend to rise and areas where the air is heavier (cooler) and tends to sink. The movements of large air masses are the cause of **wind**. But we have seen that the overall atmosphere moves in a fairly regular pattern called the **general circulation**, and that this is because parts of the globe have special characteristics: Along the **equator** there is an area of low pressure; then there are the **subtropical regions** (between latitudes 30 degrees and 35 degrees) of high-pressure, two areas of low pressure in the **temperate regions**, and two high pressure areas located at the **polar ice caps**. The air masses move among these areas of different pressure.

THE CORIOLIS EFFECT

The inertia that acts on a moving body due to the rotation of the Earth is called the Coriolis effect. The trade winds and the west wind are a product this phenomenon.

A partial view of the Earth showing large cloud currents.

→ Navigation by sail was possible because of the existence of winds on all the oceans.

For centuries the winds on the seas made maritime transportation possible; nowadays, though, they are used mostly for certain water sports.

THE GREAT WINDS THAT SWEEP THE EARTH

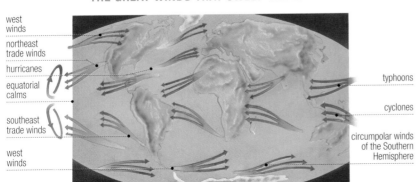

west winds
northeast trade winds
hurricanes
equatorial calms
southeast trade winds
west winds

typhoons

cyclones

circumpolar winds of the Southern Hemisphere

THE ROTATION OF THE EARTH

The **Earth** spins on its axis, producing **centrifugal** and **inertial forces** that pull the atmosphere along. In addition, **frictional forces** occur where the atmosphere contacts the ground. All these forces exert a great influence on the movement of the air masses, just as they do on the masses of water in the oceans. So when the air starts to move in response to pressure differences, the earth's rotation causes it to deviate depending on the direction of the movement: In the Northern Hemisphere the deviation is toward the right, but in the Southern Hemisphere it is toward the left. For example, the wind generated by an anticyclone in the Northern Hemisphere deviates to the right, and the resulting circulation moves around it in a clockwise direction.

CLIMATIC FACTORS I

Climate is the result of many factors that work together. Geographic features such as mountains and large oceans have a major effect on the characteristics of a climate. However, there are a few elements that are essential, and we will now have a look at temperature, humidity, and atmospheric pressure, as well as one of the most noticeable manifestations of climate, the seasons.

TEMPERATURE

The temperature is an indicator of the amount of **caloric energy** stored in the air; it is measured in degrees Fahrenheit or degrees Celsius. Temperature varies according to various factors, such as the **angle** of the sun's rays (less heat arrives at dusk when the sun is low), the type of **substrate** (rock heats up; ice reflects almost all the energy), the **wind** (the movement of the air can have a cooling effect), the **latitude** (near the poles the light always comes down at a lower angle than at the equator), and so forth. But there is a distinction between the air temperature and the **perceived temperature**, which may be lower. At 50°F (10°C) and with no wind you can be comfortable in shirtsleeves in the sun, but a 50-mi/h (80-km/h) wind produces a feeling of intense cold.

An equatorial landscape.

MICROCLIMATES

A microclimate is a set of climatic conditions that exist in a specific location and differ from the general conditions in the surroundings.

An **isotherm** is a line that joins the points of identical temperature in a region.

An **isobar** is a line that joins points at a location with the same pressure.

HUMIDITY

Humidity is the amount of water present in the air. It depends on the temperature, and warm air contains more moisture than cold air does. Various systems are used to measure humidity. **Absolute humidity** is defined as the amount of water vapor present in a specific unit of air; it is specified in grams per cubic centimeter. **Relative humidity** is expressed in percentages and refers to the proportion of water vapor present in an air mass and the maximum amount that it could hold at that temperature, in other words, if it were saturated. **Saturation** is the point at which the air can hold no more water vapor and it condenses out in the form of water.

DEW

Dew is water that is deposited on objects when the humidity in the air reaches the saturation point.

PRESSURE

Atmospheric pressure is the weight of the air mass per unit of surface. Therefore, the pressure is greater at sea level than at the top of a mountain. We can perceive the large differences in pressure because it becomes difficult for us to breathe when we climb to altitudes above 9,840 ft (3,000 m). Atmospheric pressure at sea level is about 15/lb in^2 (1,013 millibars) and it decreases progressively with altitude.

Altitude is a major factor in whether precipitation falls as rain or snow; the higher the altitude, the colder the temperature. The illustration shows the snowy summit of Mount Chimborazo in Ecuador.

THE SEASONS

Depending on latitude and altitude, the meteorological variations throughout the year can be minimal so that the same conditions always prevail (e.g., in lower equatorial regions); but in cases where the changes are major, we distinguish definite periods known as seasons. They are the most perceptible in temperate zones.

In these areas there are four seasons: spring, summer, fall, and winter. But there are variations, and so in some places there are just two seasons (one cool and moist and the other hot and dry), or one season may be much longer than the other three.

SPRING

Astronomically spring starts on March 21 and ends on June 21 in the Northern Hemisphere; in the Southern Hemisphere, it starts and ends, respectively, on September 21 and December 21. The length of the day increases, and the angle of the sun's rays changes so that temperatures rise.

SUMMER

By the calendar summer starts on June 21 and ends on September 21 in the Northern Hemisphere; in the Southern Hemisphere, it starts and ends on December 21 and March 21, respectively. Daylight lasts the longest and the sun continues in an arc high in the sky. This is the season when temperatures are the highest.

FALL

Astronomically fall starts on September 21 and ends on December 21 in the Northern Hemisphere, and on March 21 and June 21, respectively, in the Southern Hemisphere. The length of the day starts to decrease, and the sun is lower in the sky. Temperatures begin to decline.

WINTER

By the calendar winter starts on December 21 and ends on March 21 in the Northern Hemisphere, and on June 21 and September 21, respectively, in the Southern Hemisphere. This is when the days are shortest and the temperatures are lowest.

CLIMATIC FACTORS II

The three elements that we saw in the previous section are characteristic of the atmosphere. Now we will have a look at another very important climatic feature that is easily perceived: the wind, which mixes the various layers of air. To conclude, we will focus on precipitation, which, along with temperature, is commonly used in defining the various types of climates that exist on our planet.

WIND

Movement in the air is produced anytime an air mass becomes less dense in response to increasing temperature and rises to fill a void in the denser, colder air. There is a distinction between general and permanent winds that occur around the globe as a result of the **general atmospheric circulation** and other winds produced by local meteorological changes, and which may or may not be regular in occurrence.

TRADE WINDS

The permanent winds that blow toward the equator from subtropical areas are called trade winds.

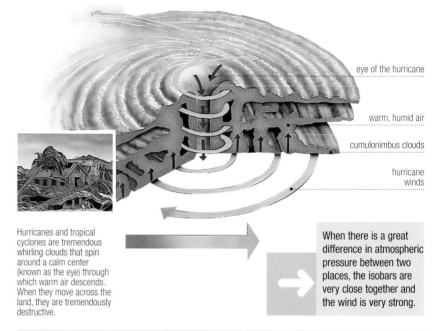

eye of the hurricane

warm, humid air

cumulonimbus clouds

hurricane winds

Hurricanes and tropical cyclones are tremendous whirling clouds that spin around a calm center (known as the eye) through which warm air descends. When they move across the land, they are tremendously destructive.

When there is a great difference in atmospheric pressure between two places, the isobars are very close together and the wind is very strong.

WIND INTENSITY CHART

Beaufort Number	Speed (mi/h)	Designation
0	0–1	calm
1	2–3	light air
2	4–6	light breeze
3	7–11	gentle breeze
4	12–17	moderate breeze
5	18–23	fresh breeze
6	24–30	strong breeze
7	31–37	near gale
8	38–45	gale
9	46–54	strong gale
10	55–63	storm
11	64–72	violent storm
12	73–82	hurricane
13	83–92	hurricane
14	93–103	hurricane
15	104–113	hurricane
16	114–125	hurricane
17	>125	hurricane

THE WEST WINDS

These are continuous winds that blow from the west in temperate latitudes.

LOCAL WINDS

In addition to the great permanent winds, the diverse topographical conditions of the planet produce winds through small regional changes. Well-known examples are the **land breezes** (fresh sea air that blows onto the land during the day), **sea breezes** (air from inland that blows onto the ocean during the night) in coastal areas, **mountain breezes** (occurring during the day and rising toward the summits), and **valley breezes** (occurring during the night and descending toward the valley in mountainous regions).

Breezes and fresh winds are ideal for sailing.

SOME FAMILIAR WINDS IN TEMPERATE REGIONS
(WHERE THEY ARE MOST COMMON)

Name	Location
Bora	Adriatic
Cierzo	Ebrus Valley
Föhn	Alps
Harmatam	Nile Valley
Poniente	Straits of Gibraltar
Simum	Sahara
Tramontana	S-France, NE-Spain

PRECIPITATION

When the humidity in the air exceeds the **saturation point**, the moisture condenses around small, floating, solid particles and forms **clouds**. When the size of the water droplets or ice particles becomes greater than the holding capacity, they fall by gravitational force in the form of **precipitation**. The cause is a cooling in the air mass that reduces its capacity for retaining water vapor, which then condenses out. This cooling off can result from mixing air masses, but in such cases precipitation is rare; it can also result from the upward movement of an air mass, and that is when large clouds form and fairly substantial precipitation falls.

The snow that builds up in the mountains is a valuable water reserve for drier seasons.

The effect of precipitation on the ground depends on the geology and vegetation; the latter serves to protect the ground from erosion. ←

Even though rain may seem to be a bother, one beneficial effect is that it cleanses the atmosphere.

TYPES OF PRECIPITATION

Depending on the temperature and the degree of condensation, there are several types of precipitation. If it falls in the form of water, it is **rain**; if it falls in crystal form, it is **snow**; and if comes down in thick ice showers, it is **hail**. When the temperature differences between two air masses are very great, the condensation can be fast and heavy, resulting in intense precipitation.

At various points on the globe there are two seasons, one dry and the other rainy; in the rainy season the precipitation can be very heavy, often causing floods.

AVERAGE ANNUAL PRECIPITATION IN CERTAIN CITIES AROUND THE GLOBE

Inches	Location
0.6	In Salah (Algeria)
1.6	Lima, Peru
2	Capo Yubi (Mauritania)
5	Verjoiansk (Siberia)
10	San Diego, U.S.
11	Coolgardie (Australia)
11	Murcia, Spain
16	Churchill, Canada
16	Prague, Czech Republic
17	Saltillo, Mexico
23	La Paz, Bolivia
28	Lisbon, Portugal
30	Reykjavik, Iceland
32	Caracas Venezuela
37	Buenos Aires, Argentina
39	St. Louis, U.S.
39	Durban, South Africa
40	Salina Cruz, Mexico
41	Gijon, Spain
41	Bogota, Colombia
42	Rio de Janeiro, Brazil
44	Quito, Equador
48	Havana, Cuba
63	Tegucigalpa, Honduras
72	Lagos, Nigeria
75	Nagasaki, Japan
75	Bergen, Norway
101	Valdivia, Chile
113	Hokitika, New Zealand
126	Sandakan, Indonesia
134	Freetown, Sierra Leone
421	Cherrapunji, India

AVERAGE ANNUAL PRECIPITATION

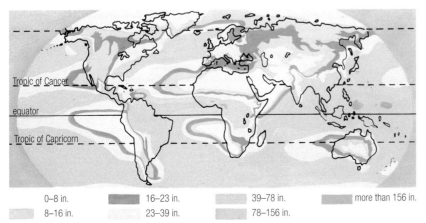

0–8 in.	16–23 in.	39–78 in.	more than 156 in.
8–16 in.	23–39 in.	78–156 in.	

PREDICTING THE WEATHER

One of the most popular sections of the newspapers and television news programs is the weather forecast. This interest is rooted in the effect that the weather has on our daily activities.

In order to predict the weather, meteorologists produce maps that show all the data, but for the ordinary person what counts is the presence or absence of clouds, and what type they are, because this information provides an indication of the approaching weather.

WEATHER MAPS

Meteorologists use the data collected at different stations to determine the atmospheric conditions in any specific area and apply laws from past experience to report how the atmosphere will change in the near future. They indicate the **cyclonic** and **anticyclonic areas**, **cloud fronts**, **isobars**, and **isotherms** on a map and use all this information to determine whether or not it is going to rain, if it is going to be windy, and the expected minimum and maximum temperatures. There are two types of maps: those made to represent **ground level** (which are used to predict the atmospheric conditions that affect us most directly) and **altitude maps** (which represent an altitude of 4 to 8 mi (6–12 km), which are used to identify the large atmospheric movements on a global scale.

SOME OF THE INSTRUMENTS IN A WEATHER STATION

Weather vane (indicates wind direction).

Hygrometer (indicates humidity in the air).

Rain gauge (indicates amount of precipitation).

Thermometer (indicates temperatures: maximum, minimum, and present).

Heliograph (indicates number of hours of sunlight).

Barometer (indicates atmospheric pressure).

A typical weather map. *H* indicates high pressure (good weather), and *L*, low pressure (bad weather).

METEOROLOGICAL INSTRUMENTS

In predicting the weather ourselves, we can use some small household instruments: a **thermometer** (for temperature), a **hygrometer** (for humidity), and a **barometer** (for atmospheric pressure). Meteorologists use many more instruments that are capable of greater precision, and, along with other factors, they also measure the wind using an **anemometer**. Weather stations are located in cloudless areas, and the instruments are placed inside ventilated cases. In addition to these traditional stations, modern meteorology depends heavily on measurements (of cloud photos, temperature records, etc.) provided by **weather satellites.**

The most precise weather predictions are for the short term— a day or two—at the local level.

Long-term predictions more than a week in advance can be made accurately only over long periods, but not for local conditions.

FRONTS, CYCLONES, AND ANTICYCLONES

When two large air masses collide, there is a sudden change in atmospheric humidity and temperature at the line of collision; this is known as a **front**. A **cold front** occurs when the advancing air is colder than the air it encounters, and a **warm front** when it is warmer than the air mass it meets. The area affected by the collision is called a **cyclone** or depression. **Anticyclones** are high-pressure areas and are the opposite of depressions.

THE MAIN TYPES OF CLOUDS

cirrus

cirrocumulus

cirrostratus

cumulonimbus

altocumulus

altostratus

cumulus

stratus

nimbostratus

stratocumulus

Introduction

The origin of the Earth

Geological history

Crystallography

Minerals

Rocks

Activity of the planet

Meteorology

Types of climate

Seas and oceans

Inland waters

Landscape formation

Erosion

Human landscapes

Cartography

Subject index

FOG

Fog is a stratified cloud that forms at ground level or close to it.

Rain and storm clouds are nimbostratus and cumulonimbus clouds.

CLOUDS

Clouds form as a result of water condensation because of a decrease in temperature. Depending on atmospheric conditions, they can take two forms and are good indicators for predicting the weather. They are classified in many ways. Depending on their shape, they may be **stratus** (flat, large, and uniform), **cumulus** (thick, with cauliflower-shaped masses that project from the top), **cirrus** (silky or filmy in appearance, and made up of ice crystals), or **nimbus** (large and dark). In addition, there are intermediate forms (cumulonimbus, stratocumulus, etc.). Clouds can also be classified according to the altitude at which they appear: **low-level clouds** (at less than 8,200 ft [2,500 m]), **midlevel clouds** (between 8,200 and 19,680 ft [2,500 and 6,000 m]), and **high-level clouds** (at over 19,680 ft [6,000 m]).

ALTITUDES AT WHICH CERTAIN CLOUDS APPEAR

Type	Altitude	Examples
high level	19,100–40,000	cirrus, cirrocumulus, cirrostratus, halo
midlevel	8,100–19,000	altocumulus, altostratus, cumulonimbus
low level	5,100–8,000	nimbostratus, cumulonimbus
low level	0–5,000	stratus, stratocumulus, cumulus, cumulonimbus

TYPES OF CLIMATES I

On a quick trip to some other part of the globe we can be surprised by a heavy storm and torrential downpours in a desert, or a prolonged drought in an area well known for its humid climate. These events are just anecdotal happenings for that climate because if we stay in that area for a longer time, we will see that it maintains well-defined characteristics corresponding to one of the climate types that we are about to discuss.

CLASSIFYING CLIMATES

We can speak in terms of **climates** or **climatic zones**. These terms refer to fairly large areas on the globe with uniform climatic conditions that differ from those of neighboring areas. Various characteristics are used in classifying climates, mainly temperature and precipitation, which are very important factors in plant growth. There are thus the following climates: **hot**, **temperate**, **cold**, and **polar** with respect to temperature, and **dry** or **wet** with respect to precipitation. Within these groups are other possible classifications depending on what is being considered. In addition, there are possible combinations among all these varieties, and so there can be cold climates that are wet or dry, climates with a single season, or with two or four, and some special climates such as **insular** and **mountain** climates.

A CLIMATE CLASSIFICATION

Climate	Type	Characteristics
humid tropical	equatorial rainy	lots of rain in all seasons
	tropical with two seasons	lots of rain but with a dry season
arid	semiarid	a short rainy season
	continental arid	very little rain
temperate humid	Mediterranean	dry summer and rainy winter
	plains	rain in all seasons
	oceanic	rain in all seasons, most in winter
cool humid	continental	limited rainfall in all seasons
	Atlantic	rain in all seasons, most in summer
polar	subarctic	rain in all seasons, most in summer
	subpolar	little rain throughout the year
	glacial	little precipitation throughout the year

DISTRIBUTION OF THE MAIN TYPES OF LANDSCAPE

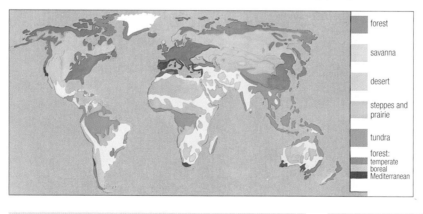

forest

savanna

desert

steppes and prairie

tundra

forest: temperate boreal Mediterranean

The African savanna during the dry season.

RAINY EQUATORIAL CLIMATE

A rainy equatorial climate forms a fringe along both sides of the equator between 10° north latitude and 5° south latitude; it is characterized by days and nights of nearly equal duration throughout the year, and there is scarcely any difference among the seasons. The average temperature is about 77°F (25°C), but there can be differences of up to 15° between day and night. The air is very humid because of the continuous, abundant rain that falls throughout the year, averaging about 78 in. (2,000 mm). The vegetation is typically that of a rain forest, as in the Amazon basin and equatorial Africa.

TROPICAL CLIMATE WITH TWO SEASONS

A tropical climate with two seasons forms a strip along both sides of the equator between 10° and 20° north latitude and 5° to 10° south latitude. There are minor variations in the days throughout the year, and the rains decrease in intensity as the distance from the equator increases (with a variation between 23 and 59 in. [600 and 1,500 mm] per year), and the average temperature is about 73°F (23°C). This means that we can speak of a dry season and a rainy season. The typical vegetation is that of a savanna, with a certain amount of forestation.

SEMIARID CLIMATE

A semiarid climate is found in both temperate and tropical areas where there is little rain, in the vicinity of 8 in. (200 mm) per year. Temperatures may be cold or hot, but there are always great daily or annual fluctuations (hot days and cold nights), especially in places where the temperatures are low. The vegetation in these areas is herbaceous and arboreal.

Canyon National Park, Utah

Because of the dryness in deserts, conditions are often very inhospitable to life.

ARID CONTINENTAL CLIMATE

An arid continental climate is found in areas designated as deserts, whose main characteristic is their extreme dryness because rain is rare, occurring only every few years. The temperatures are extreme during the course of the day because the absence of vegetation contributes to rapid heating and cooling of the ground. Deserts can be hot like the Sahara or cold like those in central Asia. However, most deserts have small areas of vegetation, especially in places where subterranean water reaches the surface and forms an oasis. One hot desert is Death Valley in California, and the driest desert is the Atacama, in Chile.

MEDITERRANEAN CLIMATE

The Mediterranean climate is typical of the areas around the Mediterranean, but it also can be found in the south of Africa, in the central part of Chile, and in California. This climate is characterized by a concentration of rain in fall and winter; in the spring and summer, it experiences a severe drought. Precipitation varies between 16 and 27 in. (400 and 700 mm) per year; the winters are mild, and the summers are hot. The usual vegetation consists of bushes and evergreen trees. There are variations within this type of climate, so it is cooler and more humid on the north shore of the Mediterranean than on the south, which comes under the direct influence of the Sahara. Farther away from the coast, this climate takes on continental characteristics and becomes cooler.

The Mediterranean coast.

TYPES OF CLIMATES II

In this chapter we will continue our description of climates because their enormous variety around the globe makes it difficult to select the most representative ones. However, we will examine those that are the most characteristic of the large groups, even though there can still be some mutual influences among them and certain areas have mixed climates.

PLAINS CLIMATE

A plains climate resembles the Mediterranean climate but with slightly higher average temperatures and less difference among the seasons. Precipitation occurs throughout the year but in greatest intensity during the summer; the average annual rainfall varies between about 23 and 31 in. (600 and 800 mm). This climate exists only in the austral hemisphere, and it is perhaps best represented by the **Argentine pampas**, although it is also found in several parts of Australia and in southern Africa. The vegetation is almost exclusively herbaceous, and there are few trees.

The Argentine Pampas. Vast plains are well suited to raising livestock.

OCEANIC CLIMATE

Landscape in Scotland.

The main characteristic of an oceanic climate is the effect of the **west winds** on the regions where this climate prevails. There are few daily or annual variations in temperature; as a result, the winters are mild and the summers are cool. The average precipitation varies between 23 and 27 in. (600 and 700 mm) per year, and in areas where there are mountain chains, these values may exceed 78 in. (2,000 mm). This type of climate is found in the coastal regions of western Europe, where the Gulf Stream mitigates the effects of latitude, and in North America, Chile, Tasmania, and New Zealand.

TEMPERATE CONTINENTAL CLIMATE

Distance from the ocean means that the temperature swings are pronounced throughout the day and between seasons. Also, the winds that arrive have lost some of their moisture and deposit scant precipitation, about 16 to 20 in. (400–500 mm) per year, evenly distributed throughout all the months. The summers are short and hot, and the winters are long and cold. This type of climate exists only in the northern hemisphere because at comparable latitudes in the southern hemisphere there are no continental masses of sufficient size. This climate is found in the interior of North America and Eurasia. The typical vegetation is that of **prairies** and **steppes,** with occasional trees.

TEMPERATE ATLANTIC CLIMATE

In a temperate Atlantic climate precipitation is moderate, with an annual average of about 30 in. (750 mm) distributed throughout the year but with a greater concentration during the summer months. In areas closest to the coast that are subject to an **oceanic** influence, higher numbers are possible. The summers are short and cool, although daytime temperatures may exceed 86°F (30°C), and the winters are long and cold. The typical vegetation is exclusively **deciduous trees** or a mixture of deciduous trees and **conifers**. There is a distinct difference among seasons. This type of climate exists in the northern hemisphere on both Atlantic shores.

Beech forests are typical of the temperate continental climate.

SUBARCTIC CLIMATE

A subarctic climate is appropriate to high continental latitudes, and one forms an extensive band in the north of Eurasia and North America. The average summer temperatures do not exceed 64°F (18°C). The summers are short, with very long days (up to 18 hrs of daylight); the winters are long, and temperatures can plummet to −76°F (−60°C) in the interior regions of Siberia. There is scant precipitation, about 15 to 23 in. (375–600 mm) per year, and it occurs mainly in the summer. The snow that falls in the winter remains for a long time in this cold. The typical vegetation is conifer forests in the form of a **taiga.**

The taiga in central Siberia.

SUBPOLAR CLIMATE

A subpolar climate exists around both poles. It is characterized by a great variation between day and night throughout the year: Daylight lasts for 24 hrs during the summer, and during the winter darkness lasts for a full 24 hrs. This factor, along with the high latitude, produces very low median temperatures that don't rise above 50°F (10°C) in the summer; in the winter, they range between −22°F (−30°C) and −40°F (−40°C). There is scant precipitation, about 10 to 12 in. (250–300 mm) per year. Because the ground stays frozen 9 months of the year and then thaws only on the surface, trees do not grow and the only type of vegetation is **tundra** consisting of mosses, ferns, and herbaceous plants.

Tundra covers a large part of the terrain in the extreme south of South America.

GLACIAL CLIMATE

A glacial climate prevails at both poles. Its characteristics are similar to those of a subpolar climate but are more pronounced, particularly with regard to temperature. A glacial summer is very short, and in most places there is barely a thaw in the top layer of the earth and at the seashore. The vegetation is reduced to a few lichens on rocks protected from the extreme climatic conditions of Antarctica and is entirely absent in the Arctic, because it is not a continent but rather an icecap floating on the ocean.

The extremely harsh conditions of the Antarctic climate mean that life is practically impossible there except for along the coastline.

55

WATER AND THE OCEAN FLOOR

The area of the globe covered by water is larger than the area of all the exposed land combined. Nevertheless, until the twentieth century the interior of this enormous liquid mass remained a mystery.

There is a very dramatic landscape underneath the sea, where new crust is continually being formed, and there are many secrets still unknown to science.

THE TERRAIN UNDER THE SEA

If we draw a line between two coasts separated by an ocean, the perpendicular plane that intersects this line reveals the four typical areas that make up the terrain beneath the sea:

1) the **continental shelf:** a plain of variable width that borders the continents and extends to a depth of about 656 ft (200 m);

2) the **continental slope:** a fairly pronounced slope that goes down to a depth of about 8,200 to 11,480 ft (2,500–3,500 m);

3) the **bottom plane:** constitutes the ocean floor at a depth of about 11,480 to 19,680 ft (3,500–6,000 m) and contains numerous mountains;

4) the **deep sea trenches:** with depths that can exceed 32,800 ft (10,000 m).

THE PARTS OF THE UNDERWATER TERRAIN

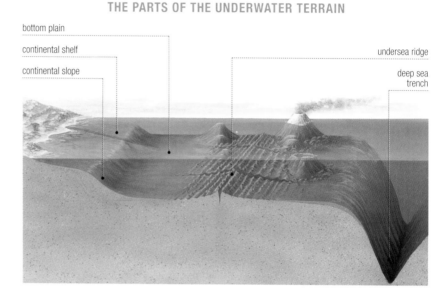

bottom plain

continental shelf

continental slope

undersea ridge

deep sea trench

UNDERWATER VALLEYS

Underwater valleys are fairly deep cuts in the continental slope caused by general erosion.

 The bottom plane accounts for 75 percent of the ocean floor.

THE OCEAN FLOOR

The **bottom plane** of the ocean floor is quite flat, and the only observable features are the great underwater mountain chains made up of **undersea ridges**, plus **sea trenches**. The ocean floor is covered with sediments that become thicker close to the continents and are scant or nonexistent at the foot of the mountains. Like mountains on dry land, these mountain chains contain basins or depressions that can be very large and are extremely important to the **ocean currents**. There are great plains in the western Mediterranean, in the Pacific south of Alaska, and along Antarctica, among other locations.

 In many places in the ocean, especially in the Pacific, the bottom is covered with spherical concretions of very pure manganese.

The continental shelf is where most underwater life exists.

UNDERSEA RIDGES

Undersea ridges are huge mountain chains that appear in the middle of the oceans. Evidently they are the result of cortex-generating activity, and they are host to intense **volcanic activity**, forming lava deposits that add to the crust. In some places these ridges have a breadth of up to 1,550 mi (2,500 km), and in others there are deep trenches that cut through and interrupt them. The series of oceanic ridges are interrelated and make up the **midocean ridge**, which is about 21,700 mi (35,000 km) long, extending through the Indian, Pacific, and Atlantic Oceans.

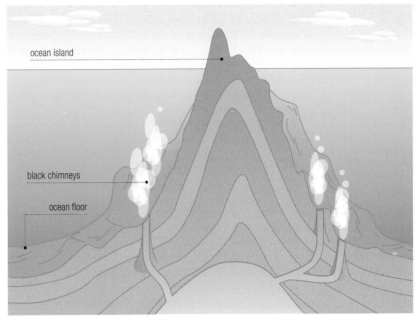

ocean island

black chimneys

ocean floor

The difficult living conditions in the depths of the oceans (where there is little food, no light, high pressure, etc.) have produced some very strange-looking fish.

DEEP-SEA TRENCHES

Deep-sea trenches are very deep areas in the ocean floor; they are always found at depths greater than 19,680 ft (6,000 m) below the surface, and they commonly are located at the edges of the oceans, or else they cut across the ridges running along the middle of the oceans. They are long and quite narrow, often with very steep sides up to 45 degrees. Their width varies between 19 and 31 mi (30 and 50 km).

THE MAIN OCEAN TRENCHES

Name	Depth (feet)
Marianas	36,152
Tonga	35,693
Kuriles-Kamchatka	34,617
Philippines	34,430
Kermadec	32,954
Chile-Peru	26,420

At some points the summit of an undersea ridge rises out of the water and forms an ocean island.

The longest known ocean trench is the Chile-Peru, which is 3,658 mi (5,900 km) long.

BLACK CHIMNEYS

Black chimneys are undersea ridges that release hot water (662°F [350°C]) that is full of black particles.

Bathyscaphs have been used to explore and study the ocean floor; the first one was made by Auguste Piccard in 1948. Later on, more maneuverable exploratory submarines replaced bathyscaphs.

Introduction

The origin of the Earth

Geological history

Crystallography

Minerals

Rocks

Activity of the planet

Meteorology

Types of climate

Seas and oceans

Inland waters

Landscape formation

Erosion

Human landscapes

Cartography

Subject index

WAVES, TIDES, AND CURRENTS

The tremendous mass of water that forms the seas and oceans is subject to different types of movement much the same way as the atmosphere is. These movements can be summarized in three large groups: waves and tides, which are perceptible on the surface, and currents, which occur within the seas and are of great significance to the climate.

WAVES

Waves are the undulations on the surface of the water that travel in a direction perpendicular to the one in which they are generated. This means that the water affected by the wave performs a **circular** or **elliptical** movement without advancing; in other words, it stays in the same place. Water located in a wave rises and falls, but it stays in place. It is the **undulations** that move. When waves reach a shallow depth such as a beach, they break and produce spray. The main cause of waves is **wind**. There are also **internal waves**, which are not visible on the surface; these are undulations in the contact area between two water masses.

How a wave works.

Many factors come into play in wave production, but wind is the main one.

Seaquakes generate vibrations that propagate as gigantic waves.

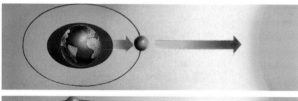

Top: spring tides caused by the position of the sun and moon; above: neap tides.

LOW AND HIGH TIDES

Low tide is the moment when the water in the tide is at its lowest and the ebb has ceased. **High tide** is the moment when the water is at its highest point.

A fishing port at low tide.

SEA CONDITIONS

Wave No.	Name	Height of Waves (ft)
0	calm	0
1	ripple	0.1–0.80
2	light swell	0.81–3
3	swell	3.1–7
4	heavy swell	7.1–13
5	heavy sea	13.1–20
6	very heavy sea	20.1–23
7	rough sea	23.1–33
8	mountainous sea	33.1–39
9	huge sea	>39

Waves generally travel at a speed of about 33 to 49 ft (10–15 m) per second.

The highest tides in the world are in the Bay of Fundy, Canada, with a difference of 102 ft (31 m).

TIDES

Tides are the regular, alternating rise and fall in the water level due to the influence of the sun and especially the moon. When the positions of the sun and moon (a new moon) coincide, both forces of attraction are added together and produce very strong tides (**spring tides**); but when they oppose one another (when the moon is full), they work against each other (**neap tides**). Tides occur two times a day, and calendars are used in ports so that boat operators will be aware of the resulting variations in water depth.

OCEAN CURRENTS

Ocean currents involve the movement of great masses of water inside the oceans and seas due to differences in **pressure**, **temperature**, and **salinity** that exist in different areas, as well as the action of **constant winds**. In the latter case, there are established, representative currents in various parts of the globe. The **Coriolis effect** caused by the Earth's rotation, which affects the direction of the winds, also has an influence on the ocean currents. These currents are extremely important to ocean life because they distribute the necessary nutrients for the growth of phytoplankton.

The nutrients carried by currents are used by phytoplankton that feed schools of fish.

OCEAN CIRCULATION

The great ocean currents are not a single mass of water, but rather are made up of many unified branches that contribute to the general movement. At first they join in two large **circuits**—the polar and the tropical—which distribute the waters in each hemisphere. But the continents cause these circuits to divide into smaller ones.

THE MAIN OCEAN CURRENTS

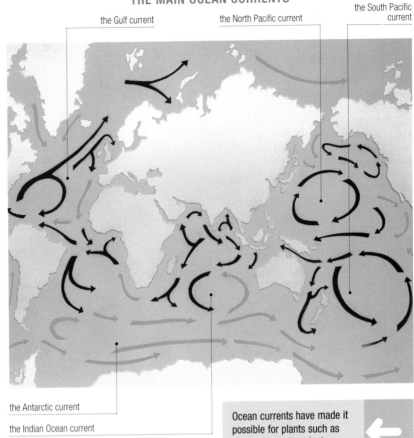

the Gulf current

the North Pacific current

the South Pacific current

the Antarctic current

the Indian Ocean current

Ocean currents have made it possible for plants such as coconut trees to colonize ocean islands.

59

SEAS AND OCEANS

The waters that fill the depressions in the Earth's crust make up the seas and oceans, which cover a major part of the planet. All together they are called the ocean, which commonly is divided into the Atlantic, Pacific, and Indian Oceans as large units; they communicate with one other through the polar oceans and seas. In turn, there are certain areas of each ocean that make up regional seas.

THE ATLANTIC

The second-largest ocean, the Atlantic stretches between Europe and Africa in its eastern reaches, and to America in its western. In antiquity the Atlantic was the limit of the known world and an insuperable barrier for navigators. The most noteworthy characteristic of its underwater terrain is the **mid-Atlantic ridge**, which runs from Iceland nearly to Antarctica. The main ocean currents are the Gulf Stream, the South Equatorial, the Labrador, the Guinea, the Canary Islands, and the Bengal. In the central area **tropical storms** are common; at the higher latitudes there are waves 66 ft (20 m) high in the winter.

The main economic activity associated with the Atlantic Ocean is fishing.

CHARACTERISTICS OF THE ATLANTIC OCEAN

surface area	41,562,300 mi^2
salinity	33–37 parts per thousand
water temperature	84°F (Caribbean) to 27°F (polar waters)

THREE CONTINENTS BORDER THE ATLANTIC OCEAN

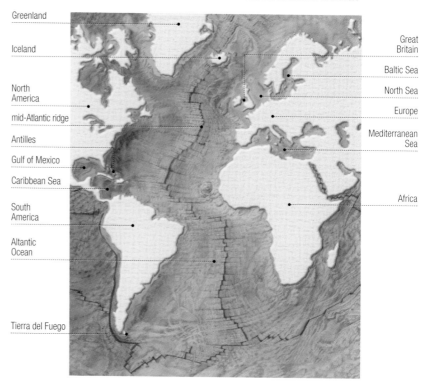

Greenland
Iceland
North America
mid-Atlantic ridge
Antilles
Gulf of Mexico
Caribbean Sea
South America
Altantic Ocean
Tierra del Fuego

Great Britain
Baltic Sea
North Sea
Europe
Mediterranean Sea
Africa

Vigo, Spain is an important Atlantic seaport.

The main oil deposits in the Atlantic are in the North Sea, in the Gulf of Mexico, and off the coast of Venezuela.

THE SEAS OF THE ATLANTIC

The main regional seas are the **Mediterranean Sea** with a surface area of 1,014,000 mi^2 (2,600,000 km^2), which separates Europe from Africa; the **North Sea** (226,200 mi^2 [580,000 km^2]), between the U.K. and the European continent; the **Baltic Sea** (163,800 mi^2 [420,000 km^2]), closed in the north by Scandinavia; and the **Caribbean Sea** (1,014,000 mi^2 [2,600,000 km^2]), between North and South America and bounded by the Antilles.

The greatest depths in the Atlantic are in the Puerto Rico trench, at 30,225 ft (9,215 m).

THE PACIFIC

The Pacific is the largest ocean on the planet; it stretches from Asia to the American continent. In its northern reaches, both continents come very close together at the narrow Bering Strait, which leads into the **glacial Arctic Ocean**, the sea by which it communicates with the Atlantic. The underwater profile of the Pacific is characterized by an expansive, flat floor in the middle and a ridge that runs along the American coastlines, bends near Antarctica, and goes as far as Australia. The Pacific contains a great number of islands but forms hardly any inner seas.

PACIFIC?

The bottom of the Pacific exhibits great volcanic activity. It is also the ocean with the greatest average depth and the one with the greatest maximum depth.

The Society Islands in French Polynesia, in the southeastern Pacific Ocean.

THE INDIAN

The Indian Ocean is the third largest ocean based on surface area and spans between the eastern shores of Africa, southern Asia, Australia, and Antarctica. It is the warmest of the three oceans, and it has the saltiest water. Its underwater terrain is characterized by a central ridge that drops down to the Arabian Peninsula and splits into two forks in the center of the ocean; one fork heads toward South Africa to join the mid-Atlantic ridge, and the other heads toward Australia, where it joins the Pacific ridge. Two of its subsidiary seas are the **Red Sea** and the sea formed by the **Persian Gulf**.

CHARACTERISTICS OF THE INDIAN OCEAN

surface area	28,906,800 mi^2
salinity	33–44 parts per thous- and (in the Red Sea)
water temperature	90°F to 27°F in polar waters

THE PACIFIC OCEAN

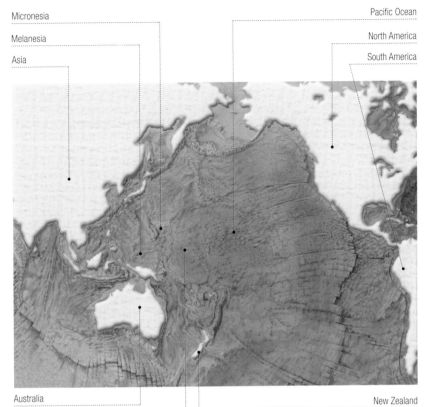

Micronesia
Melanesia
Asia
Pacific Ocean
North America
South America
Australia
Polynesia
New Zealand

CHARACTERISTICS OF THE PACIFIC OCEAN

surface area	70,722,600 mi^2
salinity	33–37 parts per thousand
water temperature	84°F to 27°F in polar waters

The main Pacific currents are the Humboldt, California, and Kuroshivo.

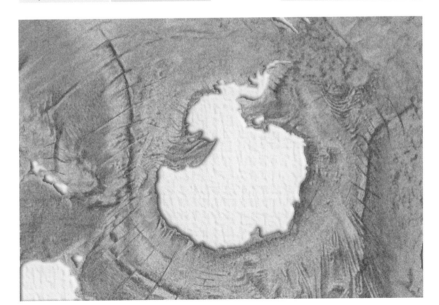

The Antarctic Sea surrounds Antarctica and serves as a meeting place for the Atlantic, Indian, and Pacific oceans.

ISLANDS

Islands can be defined as pieces of solid ground surrounded by water. They are differentiated from continents, which also appear to be surrounded by water on all sides, by their size because they are considerably smaller. They may be volcanic or coral in origin, prolongations of a continent, or the exposed tops of underwater mountains, but one characteristic they all have is that they have a climate that is strongly influenced by the ocean, as we would expect.

ISLANDS AND THEIR CLIMATE

Any point on an island is subjected to the effects of **ocean winds**; as a result, these winds are largely responsible for the humidity and the type of precipitation that prevail. If the island has **mountains**, they tend to act as screens and cause humid air to rise; the moisture then condenses and produces precipitation, and sometimes it is continual. On the side opposite the one where the wind blows, though, there is very little rain. Islands are also at the mercy of the ocean and warm currents. All this means that their climate is similar all around the world, even though **latitude** has an influence on surface temperature.

 Very small islands are known as **islets**.

ARCHIPELAGOS

Archipelagos are made up of several islands with similar characteristics that constitute a geographic unit.

The ocean has a marked effect on the climate that prevails in islands, especially if they are small in size.

The largest island in the Caribbean is Cuba, with 43,260 mi² (110,922 km²).

CONTINENTAL ISLANDS

Continental islands are located in the immediate vicinity of a **continent**, from which they are separated by a shallow strait that may emerge in certain geological ages; these islands geologically are continuations of a continent, as proven by their fossils and types of rock. One example is the **British Isles**, located on the **continental shelf** of western Europe and separated from the rest of the continent by the English Channel, which reaches maximum depths of about 100 yds (109 m).

A ferry that links Dunkirk (France) with Ramsgate (Great Britain). The British Isles are continental islands.

OCEANIC ISLANDS

Oceanic islands are located far from continents and have a different origin from them. They may appear when an underwater mountain range, or **ridge**, reaches above the surface of the water, and the exposed summits constitute islands, often with very steep terrain. At other times they are the result of huge folds or parts of the original supercontinent that did not join up with the present continents; in such cases, they can be of major proportions. Among the largest oceanic islands are **Madagascar** and **New Zealand**.

New Zealand is formed by two large oceanic islands (the north and the south). The photo shows the New Zealand Alps on the south island.

VOLCANIC ISLANDS

A volcanic island is the direct result of volcanic activity that takes place in an undersea ridge or similar structures scattered through the ocean. They often appear in huge strips and make up **volcanic belts**. These islands are still being formed, as in the case of **Surtsey**, an Icelandic island that emerged in 1963 in the course of an eruption. The Pacific has a large number of islands of this type, many of which erupt quite actively. Another example is the **Canary Islands**, which were formed during the last stages of the folding of the **mid-Atlantic ridge**.

Sometimes the growth of an underwater volcano produces a volcanic island by building up lava.

THE TEN LARGEST ISLANDS ON THE PLANET
(Australia, considered a continent, is not included)

Name	Ocean	Country	Surface Area (m²)
Greenland	Atlantic	Denmark	848,484
New Guinea	Pacific	Papua-New Guinea	308,661
Borneo	Pacific	Indonesia-Malaysia-Brunei	282,934
Madagascar	Indian	Malagasy Republic	228,944
Baffin	Arctic	Canada	197,906
Sumatra	Pacific	Indonesia	184,706
Honshu (Hondo)	Pacific	Japan	88,690
Great Britain	Atlantic	United Kingdom	85,036
Victoria	Arctic	Canada	82,757
Ellesmere	Arctic	Canada	76,532

CORAL ISLANDS

Coral islands have a biological origin. They are a result of the intense activity of **coral**, which leaves behind their calcareous skeletons when they die, forming a base structure where new coral can form. These can build up to significant depths. The deepest parts collapse and allow the upper parts to grow where there is better light. A change in sea level causes the top to emerge, and then erosion produces soil where plants can grow. These islands are quite low to the water. They also appear when the central lagoon of a circular atoll dries up. An example of this phenomenon is the islands of the **Great Barrier Reef** of Australia.

Coral islands and reefs are a product of coral life.

The largest island in South America is the Grande in Tierra del Fuego, with an area of 15,876 mi² (40,707 km²); it is shared by Chile and Argentina. It is separated from the mainland by the Beagle Channel and is home to an abundance of ocean birds.

An atoll is a circular coral reef that encloses a lagoon; one or more passageways connect the lagoon to the surrounding ocean.

SURFACE WATERS

Even though they account for only a small amount of all the water on the planet, the waters that run on the surface of exposed land are very important for all living creatures, both plants and animals. They generally are the direct result of precipitation that falls as rain, snow, and hail and the large deposits (aquifers) that this water forms; as a result of the force of gravity, many of them end up emptying into the oceans.

FORMATION

A river can arise from a **spring**; generally this water is rainwater accumulated inside the Earth's crust after being filtered through the soil, or directly from the rain and snow that fall to the ground. In many cases both phenomena happen at the same time. Some of the water that falls is filtered to form **aquifers** (deposits), and some of it carves out small furrows. Several furrows give rise to a **rivulet** that carries a greater quantity of water. Many rivulets together produce a **stream**, which may flow continuously. In any given area a number of streams can join to form a **river**, which empties into the sea if it is close by or into another, larger river if far from the coast.

A resurgence is the reappearance of water held in underground cavities.

affluent

waterfall

upper course

middle course

subaffluent

lower course

BASINS

A basin is the set of water courses that flow into and feed a main river and is separated from other basins by physical barriers such as mountains.

When a river flows through rough terrain, it forms a torrent.

RUNOFF WATER

Rainwater that runs across the surface of the ground is called rainwater.

THE PARTS OF A RIVER

When a river forms, it doesn't have much volume of flow, but its course has a steep **grade**, and so there is significant **erosion**. The bottom is made up of rocks, and there is practically no submerged vegetation. This stretch of the river is called the **upper course**. When the river has significant flow and the grade is not too steep, the speed of the current diminishes, and even though erosion still occurs, deposits still form in certain areas and underwater plants can grow. This stretch is known as the **middle course**. Finally, when there is a great volume of flow and the course flows through flat plains, the water travels at a slower speed and there is little erosion. But large deposits form, and there is significant underwater vegetation. This section is known as the **lower course**, and it ends where the river empties into the ocean.

BED—FLOW RATE

The bed is the portion of the terrain carved out by the river; the flow rate involves the amount of water that flows in the river.

THE LARGEST BASINS IN THE WORLD

River	Continent	Area (m²)
Amazon	South America	2,749,500
Congo	Africa	1,439,100
Nile	Africa	1,306,500
Mississippi-Missouri	North America	1,256,190
Plata-Parana River	South America	1,224,600
Obi	Asia	1,160,250

TYPES OF RIVERS

Rivers are commonly classified according to how they maintain their rate of flow. There are eight main types:

1. **Atlantic:** fed by rain; stable rate of flow throughout the year;

2. **continental:** fed by rain and ice melt; variable rate of flow;

3. **Mediterranean:** fed by rain; variable flow rate, often with a summer drought;

4. **polar:** fed by ice melt; little flow and only in the summer;

5. **high mountain:** fed by rain and snow, with a winter drought and a summer increase;

6. **monsoon:** fed by seasonal rains, with a summer drought and a winter increase;

7. **equatorial:** fed by rain, with two increases per year;

8. **desert:** fed by infrequent rains; scant flow and dry much of the time.

The Danube, seen here as it passes through Budapest, is the second largest river in Europe, with a length of 1,767 mi (2,850 km) and an average flow rate of 229,515 ft³ (6,500 m³) of water per second.

PRINCIPAL RIVERS OF THE WORLD

River	Continent	Length (m)
Amazon-Ucayali	South America	4,356
Nile-Kagera	Africa	4,135
Mississippi-Missouri	North America	3,979
Yangtse-Kiang	Asia	3,708
Yenisei-Angara	Asia	3,342
Parana	South America	2,914
Mekong	Asia	2,914
Amur	Asia	2,738
Congo	Africa	2,710
Lena	Asia	2,641
Mackenzie	North America	2,629
Niger	Africa	2,604
Yellow	Asia	2,573
Obi	Asia	2,492
Volga	Europe	2,290
Murray-Darling	Australia	2,164
Yukon	North America	2,040
Indo	Asia	1,978
Saint Lawrence	North America	1,947
Syr Daria	Asia	1,872
Irtysh	Asia	1,841
Brahmaputra	Asia	1,798
Colorado	North America	1,798
San Francisco	South America	1,798
Danube	Europe	1,773
Euphrates	Asia	1,711
Orinoco	South America	1,302

The highest waterfall in the world is Angel Falls in Venezuela, with a vertical drop of 3,211 ft (979 m). ←

The Guilfoss waterfall in Iceland.

The Amazon, the longest river on earth, as it passes near the Brazilian city of Manaus.

↓

When the flow rate is highest, a river is at flood stage; when it is at its lowest, drought prevails.

Introduction

The origin of the Earth

Geological history

Crystallography

Minerals

Rocks

Activity of the planet

Meteorology

Types of climate

Seas and oceans

Inland waters

Landscape formation

Erosion

Human landscapes

Cartography

Subject index

UNDERGROUND WATERS

Formerly it was believed that underground water, which was not unknown, came from the ocean and that it had lost its salinity by being filtered. In fact, we now know that this water comes from rain. It forms large deposits, which in many areas are the only source of potable water. When these waters flow beneath the ground, they sometimes carve out great systems of caves and galleries.

FORMATION

Underground waters come from rains; this has been demonstrated by calculating the total precipitation and the lesser amount of water carried off by rivers. This difference is accounted for by the amount of water that filters through the soil and ends up forming an **aquifer**. When it rains or snows on a permeable piece of ground, a significant part makes its way to lower layers until it finds an impermeable layer that stops its passage. If this impermeable area forms a depression, there is a **subterranean lake**; but if it is on a slope, there is a **subterranean river**. Permeable ground has pores large enough to allow the passage of water droplets.

FORMATION OF A SUBTERRANEAN RIVER

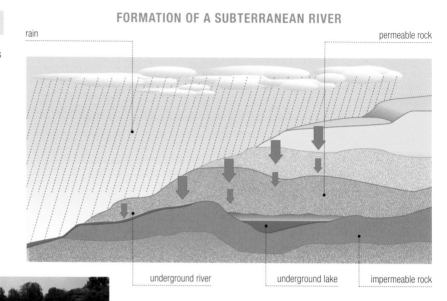

rain

permeable rock

underground river

underground lake

impermeable rock

Fractures in rock through which water seeps to refill aquifers in solid rock.

TYPES OF UNDERGROUND WATERS

Mantles can be **superficial** and appear on top of the ground; in that case they are affected by meteorological conditions and can evaporate. Other mantles are formed near rivers and are fed with rainwater; these are called **mantles of alluvial plain**. The rest can be **free**, forming a river, or **captive**, with the water compressed between two impermeable layers. When a layer becomes perforated, the water comes out under pressure.

MANTLE

The layer of underground water located under an impermeable surface is called a mantle.

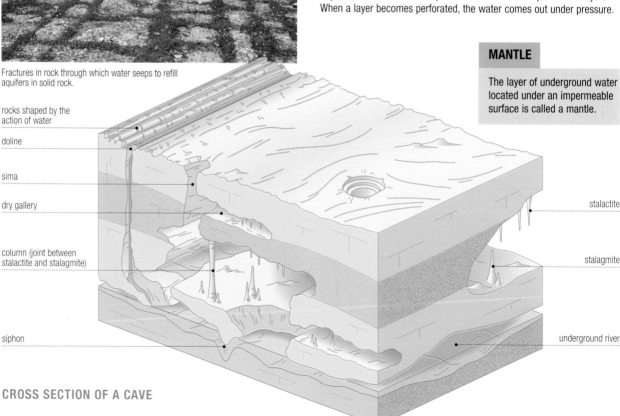

rocks shaped by the action of water

doline

sima

dry gallery

column (joint between stalactite and stalagmite)

siphon

stalactite

stalagmite

underground river

CROSS SECTION OF A CAVE

SPRINGS AND WELLS

A **spring** is a natural water outlet on the surface of the ground when the mantle is free. Springs commonly form because the impermeable layer lies on a slope that intersects with the surface at the point where the spring appears. A **well** is an artificial outlet made in a captive mantle, allowing the water to come out under high pressure whenever the well is located below the highest level of the mantle. If the perforation is located at a point higher than the mantle, the water needs to be extracted by auxiliary means, such as pumps.

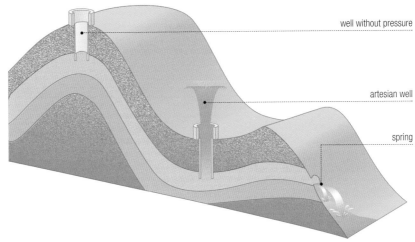

well without pressure

artesian well

spring

STALACTITES AND STALAGMITES

A stalactite is a column that hangs down from the roof of an underground cavern; a stalagmite is a column that reaches upward from the floor of an underground cavern.

 Limestone is permeable; granite is impermeable.

The geyser Old Faithful gets its name from its punctuality; it is located in Yellowstone Park in Wyoming.

Some stalactites and stalagmites develop complex and bizarre shapes; popular imagination sometimes gives them names of people, objects, or fantastic creatures. The photo shows the "demoiselle" in the Grotte des Demoiselles in southern France.

SPECIAL CASES

When unfiltered water flows through limestone, a phenomenon known as **carstification** occurs. The water dissolves the lime and leaves hollows. Eventually a large cavity or cave appears and provides a place for an **underground river**. The unfiltered water carrying lime produces column-shaped structures known as **stalactites** and **stalagmites**. At other times the water reaches deeper layers or areas where there is volcanic activity, is heated, and comes to the surface in a fountain that can shoot up dozens of yards (meters) in the form of a **geyser**.

LAKES

Sometimes lakes are considered miniature seas, and there are some similarities. They are masses of fresh or brackish water found inland on continents or islands and generally are connected to a river system.

Some lakes are an important water supply for settlements near the shore, and they are crucial to fauna.

MORE THAN WATER DEPOSITS

Unlike underground **aquifers**, which are hardly affected by climate conditions and generally have no flora and fauna, lakes are landscape features that depend on the area where they occur and frequently support considerable flora and fauna. Lakes form in natural depressions and generally fill with running water, as from a river, and also let some of the water out through a river or by filtration. However, some lakes have no outlets even though they are fed by rivers. This type of lake is common in desert areas. In other cases, the lake occupies a closed depression and receives nothing but rainwater; this is the case with many lakes located in old **volcano craters** or hollows scraped out by a **glacier**.

BASIN

The depression or cavity that contains a lake's water is called a basin.

Lake Cuicocha (Ecuador), fills the crater of an ancient volcano.

When a lake is fed by rivers but has no outlet, it forms an endoric basin.

A pond is smaller than a lake, and the body of water is always a homogeneous mix.

AREAS OF A LAKE

When a lake is deep enough, we can distinguish several different areas, much as we can in a sea. Thus, there is a **shelf** with a gentle angle near the shore; this shelf ends at a point when the **slope** increases considerably and leads to a more or less flat **bottom**. The shelf commonly contains belts of **marsh vegetation** that are very important to waterbirds. In cases where the lake is deep, there may be several layers with different densities and temperatures, and they may not mix with one another.

THE PARTS OF A LAKE

aquatic plants

shelf

bank

bottom

TYPES OF LAKES

Depending on their formation, there are several different types of lakes. It is most common for them to appear in natural depressions in the terrain (such as a fault); these lakes are referred to as **tectonic**. In other instances, when a volcano becomes extinct, rainwater collects and fills the crater, which becomes a **volcanic lake**. In the mountains many **glacial** lakes are formed after the ice disappears and the hollows in the ground fill up with water. Lakes can also form in sunken depressions when the ground dissolves, or in a crater caused by the impact of a meteorite, among other possibilities.

 Lake Baikal contains the only inland population of seals in the world.

Many of the lakes in the Pyrenees, between France and Spain, are glacial in origin.

PARTS OF A DAM

electricity transport

cornice

buttressed dam

hydroelectric plant

spillway

rock

dammed-up water

ARTIFICIAL LAKES

Throughout the twentieth century in particular, many artificial lakes have been created, and they have a major impact on their surroundings. This activity involves blocking rivers by building **dams** at narrow points. The upstream current accumulates behind the dam, often over a long distance. These dams are used to generate electric energy or as a reservoir for a community's drinking water.

 Reservoirs have some characteristics in common with rivers, and some in common with lakes.

An arch-shaped dam in the Pyrenees Mountains.

THE WORLD'S LARGEST LAKES

Name	Country	Area (m²)	Depth (feet)
Caspian	Rusia/Iran/others	14,469	3,362
Superior	United States	32,019	1,332
Victoria	Kenya/Tanzania/Uganda	27,105	269
Aral	Kazakhstan/Uzbekistan	25,155	213
Huron	Canada/United States	23,244	751
Michigan	United States	22,542	922
Tanganyika	Burundi/Tanzania/Zaire/Zambia	12,831	4,822
Bear	Canada	12,207	1,463
Baikal	Russia	11,895	5,314
Malawi	Malawi/Mozambique/Tanzania	11,271	2,280

 Half of Lake Aral, also known as the Aral Sea, has already dried up and disappeared as a result of hydraulic projects.

LAKE TITICACA

Lake Titicaca (Bolivia/Peru) has an area of 3,388 m² (8,686 km²); it is 922 ft (281 m) deep and is located at an altitude of 12,464 ft (3,800 m); it is the world's highest major lake.

GLACIERS

The huge masses of ice that cover the poles and high areas of the world's major mountain chains are called glaciers. They are the remains of the ice covering that once extended over a large part of the upper latitudes of the planet in the course of the last glaciations. Glaciers play an extremely important role in the process of erosion.

GLACIER FORMATION

All glaciers are the result of **snow** accumulation. When the depth of the fallen snow reaches a certain weight, it slowly becomes compacted. The snow crystals, which contain air, gradually lose their shape. What follows is a period of **recrystallization**, which produces a hard snow that turns into ice that's not very dense as the air is squeezed out of it. This first ice is the base of the glacier. If the depth and weight of the snow increase even further, the ice eventually compresses and turns into a very dense **ice** with no pores. When a glacier has an ice depth in excess of 22 yd (20 m), the deep layers acquire a certain degree of flexibility and the glacier becomes capable of movement.

The terminal tongue of Skaftafell glacier in Iceland.

WHITE ICE AND BLUE ICE

White ice is the fairly porous ice that appears on the surface of a glacier; blue ice is very dense, compacted ice found in the deep areas of a glacier.

Glacial ice accounts for 98 percent of the ice that exists on the planet.

A desolate landscape of arctic ice.

GLACIAL ICE

Glacial ice is the most widespread type of ice and makes up all the polar ice caps and the layer of ice that covers Greenland. It forms a continuous, very thick layer averaging about 2 mi in depth (3–4 km), and it weighs so much that in some places it has pushed the land down below sea level. At the edge of the glacier, ice breaks off and forms **icebergs**. In the Arctic, where there is no land continent, the glacial ice floats directly on the ocean; but in Antarctica, it covers nearly the whole continent, and only the peaks of a few of the highest mountains rise above it.

AN ICE RACE

The speed at which a glacier moves depends on numerous factors, such as the nature of the land, climatology, and so forth; the main ones are the steepness of the terrain and the depth of the ice. The steeper the slope, the faster the glacier moves; the thicker it is, the slower it moves. Thus, some glaciers in the Alps, such as the Sea of Ice, move about 131 yd (120 m) per year; however, some glaciers in Greenland can cover 22 yd (20 m) in a day!

ALPINE GLACIERS

Alpine glaciers are also called **valley glaciers** and commonly form in mountain areas. These glaciers arise from the accumulation of ice in high areas; when they reach a certain depth, they start to move through the effect of gravity, and they resemble a river of ice. Alpine glaciers have three main parts: The **cirque** is the area where snow falls and changes into ice; it is often more-or-less circular and limited by the sides of the mountain, except on the side where movement takes place. The **tongue of** the glacier is the river of ice, in other words, the part of the glacier that slips downhill. Finally, the area where the temperature causes the ice to melt is called the **front** or the **ablation area**. This frontal area and the sides of the glacier are where all the materials dragged along by the tongue are deposited.

THE PARTS OF AN ALPINE GLACIER

There are about 200,000 alpine glaciers distributed among the mountains of the entire planet.

cirque
compacted ice
seracs (cracks)

rimaya
lateral moraine
glacial tongue
terminal moraine
ablation area

The moraines of an ancient glacier are clearly visible beneath the peak of Mount Aneto in the Spanish Pyrenees.

Sometimes several glacial tongues join to form a huge one.

One of the most beautiful glaciers in the Andes is the Perito Moreno glacier, which breaks off spectacularly into Argentine Lake.

MORAINES

Moraines are deposits of the materials dragged along by a glacier's tongue; they can be either frontal or lateral. When the glacier recedes, the moraines may plug up a valley and give rise to a glacial lake.

THE ACTION OF AGENTS

The Earth's crust has experienced many alterations due to internal forces; it breaks up and forms again, and many of these processes are still going on. But ever since the atmosphere has existed, there have been other agents that have contributed to a slow transformation that has resulted in the Earth's present appearance. All these processes are known as erosion, and the main contributors to this action are physical, chemical, and biological in nature.

PHYSICAL AGENTS

One of the main contributors to erosion produced mechanically is **water** that freezes. When rainwater or water from any other source gets into cracks in rocks, it freezes when the **temperature** drops. Since **ice** takes up more space than water, the pressure exerted causes the crack to lengthen and can break the rock into several pieces. At other times, when the rock is porous, it first soaks up the water, and when the small droplets freeze, they expand and produce the same effect, breaking the rock into many pieces.

In very warm climates where there are great variations in temperatures on a daily basis, such as in deserts, rocks heat up considerably during the day and expand; at night, though, they cool off (even down to freezing) and contract. The quick expansion and contraction serve to break up the rock.

The breakdown of a rock due to temperature changes between day and night is more effective when the rock contains different types of minerals, each with a different rate of expansion.

EROSION

Erosion is the destruction of the earth's crust due to external agents and a carrying away of the material produced.

In desert areas, quick changes in temperature between day and night break up rocks.

CHEMICAL AGENTS

In order for chemical agents to work, the climate has to be humid because **water** is the common denominator for all these reactions. Water contains variable amounts of atmospheric **oxygen** dissolved in it; oxygen reacts with certain minerals in rocks and forms **oxides**. In other instances the oxygen and hydrogen ions that make up water attack other minerals. Every type of mineral has a different degree of sensitivity to these chemical agents. **Carbon dioxide** from the air also mixes with water and forms **carbonic acid**, which turns to water-soluble **bicarbonates** when it comes into contact with certain rocks such as limestone.

Seawater is a powerful agent of erosion; the continuous beating of waves breaks apart rocks and causes the coastline to recede.

One recent type of erosive agent is chemical air pollution. Contaminants such as sulfur dioxide and others react with water vapor and form acids.

When water from rain or snow melt encounters a certain type of terrain, it acts both as a physical and a chemical agent in erosion, producing impressive canyons such as the Arbayun Bend (Navarra, Spain).

PLANTS

Higher plants, those that have roots, perform intense mechanical **excavation** of the substrate because in their search for water they have to dig deep to find it; they are capable of boring through substrates composed of soft rocks and even breaking harder ones. Even though it is not very visible, the work of other plants and organisms, such as lichens, is perhaps even more significant. **Lichens** are considered to be the true instigators of the soil formation process because as they act on bare rock, they begin to break it down and make it possible for other organisms to continue their work.

METEORIZATION

Meteorization is a breaking down of the Earth's crust by rain, wind, etc. without the particles being carried away.

Plants that are capable of colonizing a rock and breaking it down into soil are called pioneers.

Lichens fastened to rocks hasten their decomposition.

When the roots of plants penetrate the soil, they move materials about and often contribute to erosion.

ANIMALS

Small invertebrates such as worms aerate the soil, but they also contribute to **meteorization** of the bedrock by allowing air and water to reach it, along with microorganisms that produce secretions that attack the rock. The work performed by animals generally complements that of other erosion agents with respect to the initial stages, but it becomes more important in soil formation. However, animals also produce secretions and excretions of materials that have corrosive properties; when they accumulate at specific points, they become another factor that deserves consideration.

Soil is a result of the combined action of physical, chemical, and biological erosion agents.

Small invertebrates, with their ceaseless digging, aerate the soil, transport it, and produce chemical substances; all this activity contributes to the transformation of the soil.

73

SOILS

Thanks to erosion and the activity of living organisms, the outer portion of the planet's rocky crust is converted into what we know as soil. Without it higher plants could not exist, and without them animals and humans could not live. Even though soil forms a very thin layer, it is essential for life on solid ground. Every area on the planet has characteristic soils, depending on the type of rock from which they were formed.

THE FORMATION OF SOILS

Geological, geographical, and biological factors have to come into play in order for soil to form. These include, respectively, **bedrock** and **topography**, **climate**, and **living organisms**. The type of underlying rock is important because it influences what type of soil is produced; for example, limestone rock produces a carbonated soil. Topography determines how large the layer of soil is: If it is located in mountainous terrain, the steep sides will keep it very narrow. All these **inorganic materials** are then subjected to the action of organisms, and when they are mixed with organic material, the result is **soil.** Time is also very important because the process is slow, and it takes quite a lot of time for a bare landscape to acquire a layer of soil.

Vegetation often aids in holding soil in place.

Bare soils are easily eroded.

EDAPHOLOGY

The science of edaphology studies the formation of soils and the various types of soils that exist.

Soil heats up and cools off more slowly than rock, and it retains moisture and nutrients.

PARTS OF THE SOIL

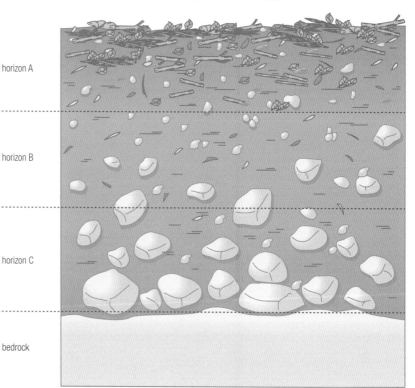

horizon A

horizon B

horizon C

bedrock

PARTS OF THE SOIL

Soil is not simply a mixture of mineral remains and organic residues, but rather this mixture appears arranged in layers. Typically a soil can consist of three layers known as **horizons**, which are designated by the letters A, B, and C. The one closest to the outside is **horizon A**, and its contents can range from totally organic material at the surface (dead leaves and organic remains) to some inorganic material in the lower part. **Horizon B** is a transition zone where part of the organic material is totally decomposed and is mixed with mineral material. **Horizon C** is the lowest layer, and the organic material gradually decreases until bedrock is reached.

AN ESSENTIAL LAYER FOR LIFE

Soil is the only medium that most plants can grow in; it contains dissolved in water the **nutrients** (salts and minerals) that plants need to carry out photosynthesis. All animals (phytophagous and carnivorous, respectively) depend directly or indirectly on plants in order to live, and many of them take shelter in the soil or live in it (worms, for instance). In addition, much of the food that humans consume is obtained directly from the soil through crop cultivation. In turn, the soil depends on the presence of plants for formation and maintenance. If the **vegetative covering** ever disappears, erosion will eliminate the soil in a short time.

Croplands not only provide food to humans but also keep the soil from eroding.

TYPES OF SOIL

Even though a soil typically contains three horizons (A, B, and C), in many cases only one horizon is present; in other soils, these horizons may be further divided into layers. This gives rise to a great variety of combinations. In addition, the soil can appear in the same place where it was formed—that is, it may be of the same type as the bedrock beneath. Other soils appear in distant locations because of the process by which soils are carried off, as by the rivers that transport mixtures of sand, mud, and organic material from upstream waters.

The cutting of forests, whether to obtain the wood for construction, fuel, or paper pulp, or to create pastures and fields, eventually causes harm to the soil.

In the mid-nineteenth century the German chemist Justus Liebig demonstrated that plants obtain their nutrients from the soil.

Between 1 and 200 years are needed to form a layer of soil 1 in. (2.5 cm) thick.

It has been calculated that about 75 billion tons of soil are lost to erosion every year.

The soil in various parts of the earth has not always looked the same. For example, thousands of years ago the Sahara was covered by abundant vegetation.

TYPES OF SOILS

Every region on the planet has a characteristic soil composition, and even though there are a great many types, they can be grouped together depending on their use or usefulness. Deserts have very expansive soils that are not suited to human settlement, in contrast to soils formed by rich sediments, which support abundant vegetation where humans can grow food and set up residence.

ARID SOILS

The quantity of **organic material** in arid soils is much reduced or totally lacking; often it is reduced to mineral material (sand, gravel, or rocks) on top of the bedrock, as in **deserts**. When conditions are a bit moister, as in predesert regions, there is a slight layer of organic material, generally only in depressions. In this instance some plants that are resistant to drought may grow. **Erosion** is very intense in these soils, especially erosion due to the wind.

In Egypt, only the soils located close to the Nile River are arable; the rest of the land is practically desert or predesert.

THE SOILS OF THE GLOBE

Type	Percent
frozen (permafrost)	6
flooded	10
fertile	11
very shallow	22
poor in nutrients	23
arid	28

There can be up to 1 million bacteria in 1 oz. (30 g) of soil, plus 100,000 yeast cells and 50,000 mycelium particles from fungi.

The soil of tropical forests is poor in nutrients and shallow, so it is not suited to long-term cultivation.

The soil of forests in temperate zones is rich in nutrients, which contribute to abundant plant growth.

FOREST SOILS

Forest soil varies considerably and depends principally on climate. Thus, in the case of **tropical rain forests**, the continual rains cause the organic material that falls to the forest floor and decomposes to be swept away quickly; as a result the amount of nutrients is very low and the fertile layer is very thin. In **temperate regions**, the change in seasons helps create a great quantity of plant residue that decomposes quickly and forms a thick layer of **humus**. This is the starting point for a soil that's rich in nutrients and very deep. In cold regions, where the **taiga** grows, the soils are acidic because of the slow decomposition of conifer needles.

AGRICULTURAL SOILS

Agricultural soil can appear in **sedimentary** terrain and in areas that have been smoothed by glaciers, where materials from other areas are deposited. These soils can be many different types depending on climate conditions. They are classified by the amount of rich nutrients in their horizons and the color of the horizons: There are **brown, red,** and **gray** soils. The layer of **humus** is thick, with a high capacity for water retention. Hardwood forests naturally grow on this type of soil, and it is a good base for natural pastures. Agricultural uses presume the removal of forests.

 One percent of the soil is organic material, and the living organisms in it make up just 0.1 percent; however, they all are crucial for fertility.

PERCENTAGE OF SOIL TYPES IN DIFFERENT PARTS OF THE GLOBE

Location	Agricultural Lands (%)	Pasture (%)	Forests (%)	Other* (%)
Africa	6	26	24	44
North/Central America	13	16	32	39
South America	7	26	54	13
North/Central Asia	10	21	32	36
Southern Asia	24	21	13	42
Southeast Asia	17	5	57	21
Australia	6	55	18	21
Europe	31	18	32	20

* Deserts, wastelands, and urban land included here.

Fertile soil is soil that is suited to plant growth (wild and cultivated); that is, it is rich in nutrients and permits root development.

Ironically, the thick vegetation of tropical rain forests occupies fragile soil.

Scarcely one-fifth of the earth's soils are suited for growing crops.

HUMUS

Humus is the layer of finely broken-down organic remains, brown to black in color, that contains nitrifying bacteria, worms, and other small invertebrates.

FLUVIAL EROSION

Inland waters are an erosive force of the highest magnitude. Rivers run across the surface or underground, erode materials in their passage, and carry the remains toward the ocean; they leave sediments in different places and make them a typical element of the landscape. Water creates waterfalls, canyons, meanders, and deltas, and at certain times of the year it causes the flooding of huge areas.

STAGES OF FLUVIAL EROSION

Erosion due to running water occurs according to the stages in which a river course is naturally divided; that is, there is a **first stage** where there is very intense **mechanical erosion** caused by the water and the materials that it carries; this occurs in the **upper course** of rivers. In the **second stage**, the transport stage, mechanical action is still important but less so than in the upper reaches; a great quantity of material is swept along, and **chemical erosion** begins to gain importance; this takes place in the **middle course** of rivers. The **third stage** occurs in the **lower course**, which is characterized primarily by sedimentation of the materials carried by the rivers; mechanical action is much reduced, and chemical erosion may still be of considerable importance.

Waterfalls are a clear example of erosion caused by water.

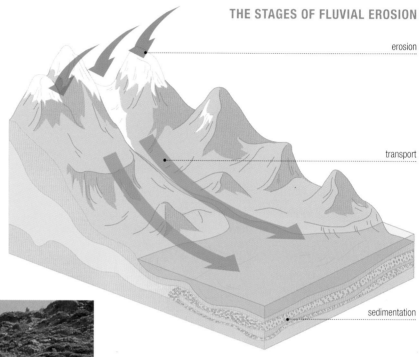

THE STAGES OF FLUVIAL EROSION

erosion

transport

sedimentation

The energy produced by a river's water is proportional to its quantity and speed.

Water movement is **laminar** when it occurs in layers that flow over one another; it is **turbulent** when these layers mix together.

THE ACTION OF WATER

Water **energy** is responsible for erosive force. It is capable of displacing pieces of rock that act like hammers on the riverbed, loosening more pieces. When the bed is irregular in shape, the water often forms **whirlpools**, which pick up sand and heavier materials; they polish the river bottom and create cavities. At other times, the steep grade causes the water to leap and act with increased erosive power. Thus, there can be very large **waterfalls** up to 3,280 ft (1,000 m) high. The edge of the waterfall gradually recedes upstream as a result of erosion. The water also dissolves materials in some rocks or attacks them and transforms them through **chemical reactions.**

FLOODS

During heavy **rainy seasons** or during a **thaw**, the flow rate of a river can increase so much that the banks cannot contain the water and it flows over the edges of the riverbed. Sometimes this phenomenon happens gradually, but at other times it is sudden and causes tremendous erosion in the whole area.

TURBID WATERS

Turbid waters contain particles in suspension because the force of the current matches the weight of the particles.

In some areas of southeast Asia, the periodic floods produce beneficial effects because they supply the necessary water to the fields used for growing rice. The photo shows flooded fields in Burma.

MEANDERS

The product of erosion is fairly fine materials that the water carries down the course of a river and that start to be deposited on the bottom in many parts of the **middle course** when the strength of the current is no longer sufficient to keep them in suspension. But erosive forces subsequently act on these deposits, wearing them away in areas where the speed of the water is greater; at the same time, new material is deposited in areas where the current is weaker. The result is deposits in a serpentine shape known as **meanders**.

A fluvial valley is not a depression but an incision in the terrain; it can, however, also occur at the bottom of a depression.

DELTAS

The end of the fluvial erosion process occurs at the mouth of the river, although in some large rivers the force of the current is capable of continuing to erode the **continental shelf**, forming an **underwater valley**. In many places, the materials the river carries down are deposited at the river's mouth, forming what is known as a **delta**; deltas are large sedimentary sections of land where there is a constant balance between the destructive force of the current and the appearance of new deposits.

The world's deltas have great biological richness, and they are the site of many national parks.

CAVES

In limestone terrain, it is common to find **underground caves** produced by chemical erosion by the water, which transforms insoluble carbonate into bicarbonate, which is soluble in water. The chemical reactions continue throughout the limestone terrain, creating large **cavities of dissolved rock**. Mechanical erosion is generally not a significant factor in this type of river.

Chemical erosion is extremely important in underground rivers.

Introduction

The origin of the Earth

Geological history

Crystallography

Minerals

Rocks

Activity of the planet

Meteorology

Types of climate

Seas and oceans

Inland waters

Landscape formation

Erosion

Human landscapes

Cartography

Subject index

WIND EROSION

In comparison to water, wind is a much less intense erosion agent, but in dry areas it takes on special importance. In these areas it forms deserts, which cover a very large area of the planet. Constant winds can give rise to such familiar features as dunes and, where they blow hardest, they can also re-form rocks into very peculiar shapes.

THE ACTION OF THE WIND

Wind by itself does not cause erosion because it does not have enough strength. What it does is transport particles that wear away the terrain upon impact. But for this to happen, it is also important for the ground to be bare because if there is any vegetation, much of the effect will be lost. Therefore, the **action of the wind** can be seen only in **deserts**. In addition, there have to be great temperature differences for the rock to break apart in the first place, and conditions have to be dry; otherwise rain or water will remove the particles from the air.

In combination with abrupt temperature changes, wind is one of the main agents of erosion in rocky deserts.

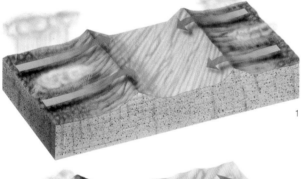

1

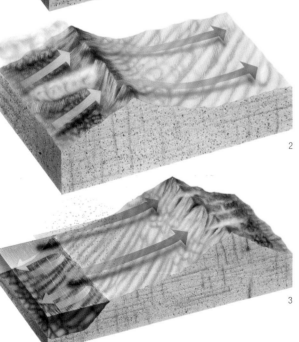

2

3

EOLIAN EROSION

Erosion caused by wind carrying mineral particles is called eolian erosion.

The deserts that experience the greatest erosive action are the hot ones, such as the Sahara.

THE FORMATION OF DESERTS

Great deserts arise in the course of a process known as **desertification**, which takes place in parts of the world where a climatic change and special topographical conditions contribute to the disappearance of **precipitation** and the erosive action of the **wind.** The changes in the **general circulation** of the atmosphere mean that winds loaded with moisture from the oceans, the usual means of delivering moisture inland, do not reach that area. In other cases, as in Atacama, Chile, even though the ocean is close, a very high mountain chain blocks the humid winds. They lose their moisture during the ascent, and when they go down the other side, they are totally dry. The absence of water inhibits **plant growth**, and this in turn favors the erosive power of wind.

Three possible situations produce deserts:
1. Inland deserts are far from the ocean, and often mountain chains form barriers to rain.
2. When a moist wind hits a mountain, it rises and cools off, depositing its water in the form of rain that is then unavailable for the other side of the mountain.
3. Coastal deserts are the product of cold air currents that reduce marine evaporation and carry very little moisture.

DUNES

When the wind reaches a certain force, it is capable of carrying along particles of various sizes. The largest ones fall back to earth when the wind speed decreases, but the finest ones continue moving until the wind dies down and they settle out. In this way, there is a stratification of the materials produced and transported during **eolian erosion**. The lightest particles end up accumulating in large deposits on which the wind acts to form **dunes**. These are accumulations of fine sand, commonly in a half-moon shape with a gentle slope on the windward side and a steep grade on the leeward side. The top of a dune erodes because of the wind, and the material moves toward the front, so the dune can move if it encounters no obstacles.

Dunes are mountains of sand that form around an obstacle as a result of the action of wind. The photo shows the dunes of Rub al-khali on the Arabian peninsula.

SOME OF THE PRINCIPAL DESERTS OF THE PLANET

Name	Continent	Location	Area (m²)
Sahara	Africa	northern half	3,549,000
Libya	Africa	NE of the Sahara	655,200
Australian	Australia	central area	604,500
Arabian	Asia	Arabian peninsula	507,000
Gobi	Asia	central	405,600
Sonora	America	Mexico	120,900
Kalahari	Africa	Botswana	101,400
Thar	Asia	India–Pakistan	101,400
Atacama	America	Chile	70,200
Namibian	Africa	Namibia	63,180
Mojave	America	Mexico–United States	25,350
Negev	Asia	south of Israel	4,992

COASTAL DUNES

Along the coasts of many seas there are dunes on the **beaches**. The process of their formation is similar to what happens in deserts. The constant breezes that blow in these areas often form large areas of **movable dunes** that can cover up vegetation; however, when the dunes are held in place, it is the plants that succeed in colonizing them.

HAMADA

A rocky desert that has no dunes is called a hamada.

ERG

In the Sahara, a large field of dunes is known as an erg.

Coastal dunes in Gata Cape, Almeria, Spain, next to the Mediterranean.

GLACIAL EROSION

Glaciers are major factors in erosion; as they passed through, they shaped a great part of the present landscape in the middle and higher latitudes of the entire planet. The tremendous masses of moving ice slowly wear away the land over which they slide, and the effect is still visible in areas where glaciers have disappeared.

HOW GLACIERS FUNCTION

In contrast to what happens with **river** water, the functioning of a **glacier** is still not fully understood. Scientists have proposed several hypotheses, but there are still many unanswered questions. In any case, there are two main ways in which glaciers work. One involves **mechanical action**. The **ice** creates great pressure because of its weight, and it wears down the rocks in the channel along which it moves. In addition, it drags along rocks of all sizes, and as they move, they also wear on the walls and accumulate at the end of the glacier to form **moraines**. Also, the water from ice melt performs a very important **chemical function** by helping to break down the rock, a process that is completed by the movement of the ice mass.

The joining of two glaciers. The lateral moraine of each one then becomes a central moraine.

A snowfield is the area where snow accumulates; with time and more snow, it can turn into the cirque of a future glacier.

RESULTS OF THE PASSAGE OF A GLACIER

When a glacier moves down a mountainside, it **carves out a valley** that commonly is U-shaped. The walls show striations in the softest areas of rock from the passage of materials carried along. Rubble in the form of larger rocks accumulates at the bottom and is more abundant at the end of the glacier's tongue, where **moraines** are formed. Sometimes large, isolated masses of rock are found on old glacial terrain; these are called **erratic blocks**, and they were caused by the force of the ice.

THE LIMIT OF PERPETUAL SNOWS

This is the altitude at which fallen snow can remain for the entire year without melting and can give rise to a glacier.

Erosion on the walls enlarges the cirque of a glacier; it may join another glacier if the wall that separates them is worn away.

Snowdrifts are corners or depressions in the upper reaches of mountains where snow and ice remain for a long time.

THE LIMIT OF PERPETUAL SNOWS IN DIFFERENT AREAS

Location	Altitude (feet)
Antartica	0
Spitzbergen	1,640
Tierra del Fuego	1,640
Greenland	3,280
Norway	4,920
Alps	9,512
Aconcagua	12,136
Caucasus	12,464
Peru	14,760
Himalayas	16,400
Kilimanjaro	17,056
Bolivia	18,040

AVALANCHES

One form of erosion different from that caused by **glaciers** but still due to snow accumulation is the erosion caused by **avalanches**. Avalanches are large masses of snow that break loose from higher altitudes and tumble down a valley, dragging along vegetation and rocky material. They leave a swath of bare ground where **erosion** can have an even greater effect. Avalanches occur when a great amount of snow collects in a steep area, especially when it is on top of a layer with a different consistency that sets up a very delicate balance or allows it to slide. Through gravity, or in response to some vibration such as the passage of skiers or a loud noise, the snow mass starts to slide, and as it goes down the mountain it picks up more snow until it carries a tremendous mass.

Avalanches do not just ravage the land: They often produce property damage and cause deaths.

Avalanches are more frequent in the spring and following periods of rising temperature.

POLAR GLACIERS

The icecaps that make up the **glacial ice** in the Arctic and Antarctica are another type of glacier of even greater volume. They behave differently because generally they cover flat ground or float on water, as in the Arctic. In Antarctica, the glaciers in mountainous regions function in the same way as alpine glaciers, but they are fused with a great mass of glacial ice. The ice moves along and exerts enormous pressure on the bottom, dragging along pieces of rock that are subsequently deposited on the ocean floor near the end of the glacier. During the glaciations, polar-type glaciers covered a large part of North America and Europe and left some distinctive contours.

In the last several thousand years, glaciers have been receding and have revealed their effects: U-shaped valleys, fragmented and polished rocks, deposits of transported materials, new lakes, and so forth. The photo shows Grey glacier in the Chilean Patagonia.

DRUMLINS

Hills with gently sloping sides separated by depressions; they were formed during the passage of ice during the glaciations.

Icebergs (a word of Flemish origin meaning "mountain of ice") are large blocks of ice that have broken off the continental ice masses and float and move around in the ocean. They are a grave danger to ocean travel.

MARINE EROSION AND THE COASTS

The coast is the area that marks the boundary between solid ground and the ocean; it is subjected to the strong erosive force of the water, and so it can form in very different shapes depending on the type of terrain and the activity of the waves and ocean currents. There are cliffs and beaches and deltas and estuaries, and sometimes a coast is cut into an ancient, flooded valley.

THE FORCE OF THE SEA

By virtue of their mass, the oceans have tremendous mechanical power, which constitutes the major **erosive agent** that shapes coasts. This action is carried out in different ways by the ocean's activity. One involves the action of the waves, which continually beat against the shore; in addition to their own inertia, the **waves** also carry sand and even rocks, which greatly erode the geological structure of the coast. Another significant action involves the **ocean currents**, which act permanently on the materials deposited by the waves. Finally, the **tides**, as they raise and lower the sea level, carry many materials that erode the bottom. In addition to these mechanical actions, there is the **chemical effect** of the water on the rocks, and to a lesser extent, the erosive action of the wind.

PARTS OF THE COAST

- cliff
- abrasion shelf
- beach
- erosion deposits

LITTORAL EROSION

The shaping of coasts caused by waves, currents, and tides is called littoral erosion.

The entire area uncovered around the world at low tide is about 58,500 mi^2 (150,000 km^2).

PENINSULAS

A peninsula is a fairly large portion of land surrounded by water on all sides except for one fairly narrow one known as an isthmus, which connects it to the larger land mass.

Beaches are formed by the accumulation of sand or gravel.

CLIFFS AND BEACHES

When exposed land rises up abruptly, the littoral line commonly forms a **cliff**, a high, steep, even vertical area that can reach significant altitude. The action of the **waves** and the **currents** removes rocky material, which accumulates at the foot of the cliffs. It forms a deposit that initially remains underwater, but which may eventually emerge to form a small beach known as an **abrasion platform**. In gentle terrain, the action of the waves produces **beaches**, which are accumulations of fine, rocky material (sand). This **accumulation** of sand extends below the surface of the water and blends with the material on the bottom. **Dunes** often form on the exposed areas of beaches.

CAPES AND GULFS

There are many geographical features that form between the ocean and the coast. Thus, a **cape** is a part of the coast that juts out sharply into the ocean; in contrast, a **gulf** is where a large, curving section of ocean penetrates into the coast, and at each end there is commonly a cape; a **bay** is like a gulf, but it is more open and smaller in size; a **cove** is a small, protected inlet, sometimes with steep sides.

When waves beat directly on its base, a cliff is said to be **live**; however, when there is an area of deposited material, the cliff is considered **dead**.

A coast where the main feature is cliffs.

Underwater vegetation and bottom-dwelling animals also contribute to shaping the coastline, especially in the presence of coral.

ESTUARIES, DELTAS, AND COASTAL MARSHES

Marine erosion is not limited to the coastline, for it can also penetrate a certain distance upstream in rivers. The area of a river's flow where the effect of the tides is perceptible is known as an **estuary**. The sedimentary materials that a river carries along arrive at the sea and are deposited near the mouth; the action of the ocean works against the force of the river, and the result is a **delta**, which is subjected to the process of erosion. These competing forces mean that a delta may frequently change appearance and new beaches and intermediate wetlands may be formed. Sometimes deposits of sand accumulate parallel to the coast and end up closing off a portion of seawater, creating a **coastal marsh**, which can turn into a **swamp**. Coastal marshes also appear when branches of a river in a delta are closed off and become subject to the action of currents and tides.

Swamps are wetlands near the ocean. Sometimes they are used for raising livestock or for growing crops. They are also important resting areas for migratory birds. The photo shows some of the swamps of the Camargue in France.

OTHER TYPES OF COASTS

In addition to the above-mentioned types of coasts, another type of coast is formed by ancient valleys on land that have become flooded by seawater. They then become rias and fjords. In such cases the coasts have a shape that can be very steep. When the valley was fluvial in nature and it contained the river that shaped it, the coast is said to be a **ria**. When the valley was part of an ancient glacier, the coast is termed a **fjord**.

Fjords are present in more highly evolved form on the western coasts of all continents.

A ria is a fluvial valley invaded by the ocean and influenced by the tides.

HUMAN LANDSCAPES: EXPLOITATION

To a certain extent all organisms alter the environment that surrounds them, and as we have already seen, some of them are important agents in shaping the landscape. Humans have a particularly high capacity for altering their environment because of their technical advances. This means that landscapes that have arisen through human intervention occupy a considerable amount of the planet's surface.

AGRICULTURAL LANDSCAPES

Ever since the invention of **agriculture**, humans have continued to occupy more and more land for cultivating plants to feed themselves or their animals. In primitive settlements, such as those of some Amazon tribes, a small section of forest is cleared for growing plants as long as the ground remains fertile. After a few years this plot has to be abandoned and another area has to be cleared for cultivation. **Traditional agriculture** is more advanced than this; it sets aside land permanently for cultivation and uses **fertilizers** to maintain the land's fertility. The plots commonly are separated by natural hedges and small woodlots. Finally, **industrial agriculture** sets aside huge areas for growing the same plant; in other words, it practices **monoculture**.

Monoculture favors the appearance of diseases and annihilates the indigenous flora and fauna.

Lands under traditional cultivation, with natural hedges or woods, permit the existence of native fauna.

Cultivation radically changes the ground.

FOREST LANDSCAPES

Wood is necessary in construction, for fuel, and for making paper. But because many forests have been disappearing, many countries have set aside large areas of land for **forest crops** of important species. These forests are in fact **monocultures**, and in general they have a negative impact on the flora and fauna of the area. **Conifers** are some of the main forest species used for these purposes. Eucalyptus, poplar, and other trees are also grown. The careful management of a forest area can prevent damage to natural forests.

Plans for cutting and replanting a forest should maintain the native species and avoid replacing them by exotic species that grow more quickly or have greater commercial potential.

Unrestricted cutting of valuable tropical forests is causing their disappearance.

LIVESTOCK LANDSCAPES

Livestock landscapes include **fields** and **pastures** where forage is grown for **domestic animals**. There are natural areas such as the North American plains and the South American pampas; in the course of the twentieth century these areas have been converted to cattle ranches, with fences and installations for storing fodder and taking care of the animals. Livestock landscapes are an important alteration for the forest fauna because it cuts down on their resources.

Cutting trees in Amazonia to make pastures for cattle is one of the factors in the destruction of the rain forest.

A pasture in Scotland.

MINING

Strip mining, or mining in the open air, causes a major alteration to the countryside. The tailings from this type of activity can destroy mountains and change the course of rivers and aquifers. Underground mining by digging galleries does not appear to alter the landscape, but the auxiliary installations and their residues sometimes cause serious damage.

A quarry is an open mine used for mineral extraction (for stonework, clay, etc.). These mines completely destroy the beauty of the landscape and leave the bedrock exposed.

The mining of metals has been carried out for several thousand years. Mining for energy materials such as coal, however, is a fairly recent phenomenon.

DAMS AND RESERVOIRS

Beavers have been building small **dams** for thousands of years, but the dams that people construct have a great potential for altering the landscape. Some of them are more than 109 yd (100 m) high, and others are hundreds of yards (meters) long. The result is a tremendous capacity for storing water; dams are used for storing **drinking water** for cities, **irrigating** crops, and producing **electrical energy**.

The great Aswan Dam in Egypt provides a large percentage of the electrical energy that the country needs and controls the floods of the Nile; however, it is causing significant ecological damage.

HUMAN LANDSCAPES: CONSTRUCTION

Pastures, agricultural lands, and open pit mines are alterations created for extracting natural resources, whether for food or for energy. Other alterations to the landscape designed for human activities generally cause greater disturbances and generate residues that often are pollutants. We will now examine some of them.

URBAN LANDSCAPES

There have been **cities** ever since ancient times, but their growth has accelerated since the industrial age, and nowadays in many countries most people live in **urban centers**. Large **metropolises** completely alter the environment and create a unique microclimate around them. Urban landscapes are characterized by great resource consumption (drinking water, electricity, food, energy, etc.). In addition to the environmental problems, urbanization requires space for construction that generally is detrimental to agricultural lands.

The lighting used in large cities is so intense at night that it can be seen from outer space.

The average temperature of large cities is always a few degrees higher than that of the surrounding area.

PORTS AND AIRPORTS

Ports are infrastructures constructed on coasts; they can seriously alter the flow of marine currents, changing the natural balance of littoral ecosystems. Although passenger transportation has a lesser impact nowadays, the shipment of goods and energy products such as petroleum makes ports strategic installations in all countries. Around these ports there has to be a large network of **auxiliary installations** (warehouses, fuel reservoirs, links to other communication means, etc.). **Airports** are located near large cities, and passenger traffic there is as important or more important than commerce in merchandise.

The main problems associated with air traffic are air pollution and noise.

TRANSPORTATION ROUTES

Transportation is an essential element in all modern societies, but it exacts a tremendous toll on the natural environment. Unless they are constructed according to environmental norms, **highways**, **roads**, and **railroads** divide natural landscapes into closed units, and this is particularly detrimental to forest animals. These transportation systems also require the construction of bridges and tunnels, along with auxiliary facilities such as gas stations, restaurants, railroad stations, and so forth.

In order to allow animals freedom to move, some railroad beds for fast trains and some modern highways have passages designed for them.

The road system of every large city requires an infrastructure that has a significant impact on the landscape.

 Minerals are resources that can be recycled, but energy products such as coal and petroleum are nonrenewable resources.

VISUAL POLLUTION

For several years now the concept of visual pollution has been used in referring to construction that may not harm the landscape or the soil but rather destroys its beauty. This is what happens when a high-tension line passes through a particularly beautiful valley, billboards are ereted next to highways, monuments and symbols are placed on mountain peaks, and so forth.

INDUSTRIAL CENTERS

Industrial landscapes are one feature that has acquired a certain notoriety since the nineteenth century. Although they are production centers for certain **goods**, they are also recipients of great quantities of **resources**. In addition, they are often located in areas that have good transportation facilities. The main problem that factories and energy plants cause is **pollution**; even though industrialized countries have succeeded in reducing pollution, it is still a serious environmental problem. However, ever since the second half of the twentieth century, the worst industrial polluters have been moving to less developed countries where there are fewer restrictions.

Industrial and manufacturing complexes take up large areas and pollute the air and water.

REPRESENTATIONS OF THE EARTH

People have always had a need to move from one place to another—sometimes in search of food or new territory and at other times to reach distant markets for their commercial activities or to conquer enemy cities. In order to do those things, they needed to know the shape of the Earth. We will now look at some of the techniques that have been developed so that people could know just what our planet is like.

THE SHAPE OF THE EARTH

In antiquity people were interested in knowing about the lands that they and their neighbors inhabited. Most movement took place within known areas. But interest kept growing in what the **shape** of the Earth was. Evidently the Egyptians thought it was round, as did **Pythagoras** and **Aristotle**, although the Greek world thought it was a flat plane floating in space. This idea lasted to some extent until the explorers of the sixteenth century circumnavigated the globe with their ships.

Old maps were drawn based on annotations made by explorers; these were in turn used as guides for new explorations.

GEODESY

The science that studies the shape of our planet is called geodesy.

DEGREE

A degree is one of the 360 units into which a circle can be divided.

A REPRESENTATION OF THE TERRESTRIAL GLOBE

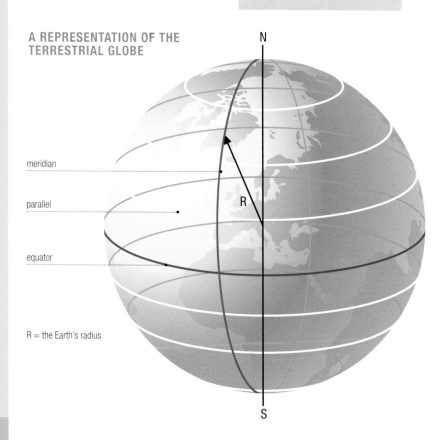

meridian

parallel

equator

R = the Earth's radius

THE SURFACE AREA OF THE PLANET

Eratosthenes (276–195 B.C.) was the first to calculate the **radius** of the Earth's sphere. This measurement is essential in determining all others that are made on the surface. The initial belief was that the Earth was a sphere, but subsequent calculations, such as those of Isaac Newton, demonstrated that this sphere was flattened at the poles. With advances in mathematics, these calculations were perfected, and in the mid-nineteenth century an international project was begun to perform precise measurements in many areas in order to establish what constituted a **degree**. After 20 years, scientists had enough data to create a very precise representation of the planet. It is a mathematical model known as a **geoid**, but the real surface has slight variations and in some places it is a little more flattened than in others.

 The differences in gravity between various parts of the Earth's surface indicate that there are irregularities in this surface.

Introduction

The origin
of the Earth

Geological
history

Crystallography

Minerals

Rocks

Activity of
the planet

Meteorology

Types of
climate

Seas and
oceans

Inland
waters

Landscape
formation

Erosion

Human
landscapes

Cartography

Subject index

TOPOGRAPHIC INSTRUMENTS

Nowadays many different instruments are used in making measurements on the surface of the Earth. The simplest ones are a **plumb line** and a stake with marks located at the distance that needs to be calculated. A **theodolite** is used to determine the distances and angles between different points. These methods are still used to find distances and heights on land for such purposes as constructing a highway or a building. For measurements of greater magnitude, very precise instruments are used, and people also make use of **gravity** meters, **radio** waves, and, recently, **artificial satellites** that make it possible to calculate very precisely the distance between any two points on Earth.

A measuring tape is a band of reinforced fabric divided into units. It is used to measure small distances.

A theodolite is a precision instrument for measuring angles; it has lenses used to locate the reference points.

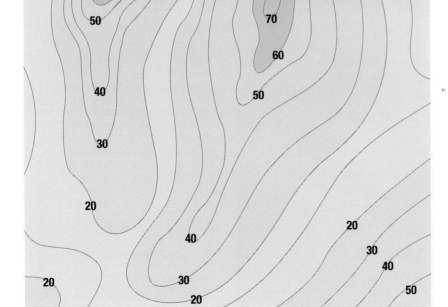

METHODS OF MEASURING

For measuring short distances, direct means are used, such as a **measuring tape**. For other distances, indirect, geometric methods are used: The **angles** of a triangle are determined when the length of one of the sides is known and the length of the others can be calculated. This method is called **triangulation**. Successive triangles are constructed on the surface of the land, and all the distances are calculated. It is also necessary to calculate the **altitude** of each point with respect to sea level. This way a series of **spot heights** is produced, and when all the similar ones are joined on a map, they produce a closed line called a **contour line**.

SPOT HEIGHT

The altitude of a point on the surface of the Earth relative to sea level is called the spot height.

Contour lines make it possible to know the shape of a specific piece of land. The closer together the lines are, the steeper the grade at that point. We can make out two hills in the illustration; the one on the left is 262 ft (80 m) high, and the one on the right is 230 ft (70 m) high.

REPRESENTATIONS OF THE EARTH: MAPS

Some of the most ancient maps known are pieces of leather on which simple marks have been traced. Today we have all kinds of maps suited for the most diverse activities. The problems associated with representing the Earth's surface on a flat plane have been solved using the techniques presented in the preceding section.

SYSTEMS OF REPRESENTATION

It is possible to prepare maps with the help of data obtained from topography. The main problem is that we have to represent on a flat surface something that in fact is a spherical surface. Various ways to solve this problem have been tried since ancient times. Presently **geographic projections** are used; in other words, the topographical data are changed to values on a flat surface, and for that purpose, small corrections are necessary. The Earth's surface is divided up into sections known as **geographic reticula**.

CARTOGRAPHY

Cartography is the set of techniques used in making maps of the Earth's surface.

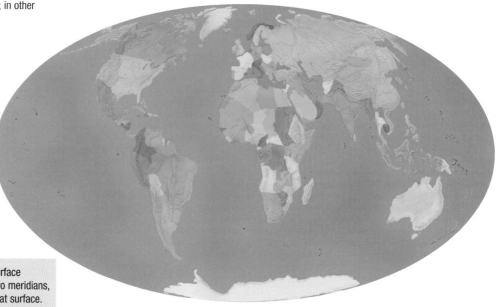

World maps make it possible for us to see the entire surface of the Earth, but the shapes of geographical features are distorted.

A GEOGRAPHIC RETICULUM

This is a section of the Earth's surface delimited by two parallels and two meridians, which can be represented on a flat surface.

ARTIFICIAL SATELLITES

Ever since the first satellites were launched into space, they have been used in making increasingly precise maps of our planet's surface. Satellites can be used to measure distances with the aid of radio emissions, and they can also take very precise photographs. A satellite takes photos of small sections of the Earth's surface, and then they are put together to make a precise representation of the shapes of the geographical features. These methods have finally made it possible to create a true image of our planet.

A segment of the Earth (specifically North and Central America) photographed by an orbiting satellite.

TYPES OF MAPS

There are many types of maps, each of which is used for a specific purpose. Depending on the intended use, the representation contains a certain degree of detail. For a **world map** of the entire planet, there is less detail, and many geographical features are even left out. The scale of the map indicates the degree of precision involved. The following types of maps are distinguished, depending on their scale:

1) **geographic maps** (on a scale of 1:1,000,000 or greater) for representing the continents or the entire globe;

2) **orographic maps** (on a scale of 1:100,000 to 1:1,000,000) for showing in detail the major geographical features;

3) **topographic maps** (on a scale of 1:10,000 to 1:100,000) for faithful representation of the terrain and artificial obstacles; and

4) **cadastral**, **urban**, and other maps (on a scale of less than 1:10,000) for representing the precise limits between properties and for other such purposes.

SPECIAL MAPS

Starting with any one of the major types of maps, it is possible to make other specialized maps designed for specific purposes. A good example is a **road map**, which represents on a flat surface the location of roads and towns. These maps commonly do not show geographical features, except those that directly bear on the roads, such as rivers that have to be crossed.

Hiking or **trail maps** represent in detail and on a small scale the natural and artificial

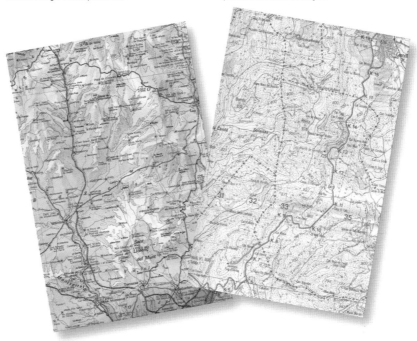

Geographic map (scale of 1:25,000). The smaller the scale the larger the map detail is.

Topographic map (scale of 1:50,000). The contour lines provide information on heights.

features (such as rivers and lakes, or buildings and bridges, respectively) and provide an idea of the relief using some contour lines. The closer these are together, the steeper the grade, and so these maps are very useful in planning a travel route. These maps also indicate the orientation of the main directions.

SCALE

Scale is the relation between the dimensions of a map and the actual distances. So a scale of 1:10,000 indicates that 1 cm on the map equals 100 m on the ground.

Geologic map. Different colors are used to indicate the various components of the soil.

Road map. The most important feature is the roads.

In order to use a trail guide correctly, you will also need to use a compass.

SUBJECT INDEX